Synergy in Polyphase Materials: Harnessing the Power of Glass and Ceramics

Synergy in Polyphase Materials: Harnessing the Power of Glass and Ceramics

Guest Editors

Michael Ojovan
Georgiy Shakhgildyan
Kai Xu

MDPI

Basel • Beijing • Wuhan • Barcelona • Belgrade • Novi Sad • Cluj • Manchester

Guest Editors
Michael Ojovan
School of Chemical, Materials and Biological Engineering
University of Sheffield
Sheffield
UK

Georgiy Shakhgildyan
Department of Glass and Glass-Ceramics
Mendeleev University of Chemical Technology of Russia
Moscow
Russia

Kai Xu
State Key Laboratory of Silicate Materials for Architectures
Wuhan University of Technology
Wuhan
China

Editorial Office
MDPI AG
Grosspeteranlage 5
4052 Basel, Switzerland

This is a reprint of the Special Issue, published open access by the journal *Materials* (ISSN 1996-1944), freely accessible at: https://www.mdpi.com/journal/materials/special_issues/07TMCV7U9S.

For citation purposes, cite each article independently as indicated on the article page online and as indicated below:

Lastname, A.A.; Lastname, B.B. Article Title. *Journal Name* **Year**, *Volume Number*, Page Range.

ISBN 978-3-7258-7603-7 (Hbk)
ISBN 978-3-7258-7604-4 (PDF)
https://doi.org/10.3390/books978-3-7258-7604-4

Cover image courtesy of Natalia Klimenko
Glass melting process

Contents

About the Editors

Michael Ojovan

Michael Ojovan, Wuhan University of Technology, is a leading authority on radioactive waste management and glass transition science. He has held senior roles at Imperial College London, the International Atomic Energy Agency, Tohoku University, and the University of Sheffield, and currently serves as Chief Editor of *Science and Technology of Nuclear Installations*. Author of 16 books, he is recognized for pioneering theories and materials that advance nuclear waste immobilisation and for innovative concepts like self sinking hot capsules to study Earth's deep interior.

Georgiy Shakhgildyan

Georgiy Shakhgildyan is an Associate Professor at D. Mendeleev University of Chemical Technology of Russia, where he obtained his Ph.D. in Chemical Engineering. His research focuses on the design and characterization of functional oxide glasses and glass—ceramics, with a particular emphasis on structure–property relationships, phase separation and crystallization phenomena, and applications in photonics and optoelectronics. He has authored more than 70 peer-reviewed publications in leading international journals and holds multiple patents related to glass composition and processing technologies. His work has been supported by several state-funded research grants, and he is a recipient of the Moscow Government Award for Young Scientists. In addition to academic research, he actively contributes to interdisciplinary projects at the interface of glass science, functional ceramics, and applied materials engineering.

Kai Xu

Kai Xu has been a professor with the State Key Laboratory of Silicate Materials for Architectures at Wuhan University of Technology, China since 2016. He is currently serving as the chair of the Technical Committee of Waste Vitrification (TC-05) on the International Commission on Glass (ICG) and the academic editor with the journals of *Science and Technology of Nuclear Installations* as well as *Research & Design of Nuclear Engineering*. He has expertise in glass-based materials and hazardous waste immobilization. His current research interests are on vitrification of hazardous and nuclear wastes including waste glass composition–property correlation, the batch-to-glass transition, precipitation and crystallization of waste glass, as well as numerical simulation of the melter.

materials

MDPI

Editorial

Synergy in Polyphase Materials—Harnessing the Power of Glass and Ceramics

Georgiy Shakhgildyan [1], Kai Xu [2] and Michael I. Ojovan [3,*]

1 Department of Glass and Glass-Ceramics, Mendeleev University of Chemical Technology, 121165 Moscow, Russia; shakhgildian.g.i@muctr.ru
2 State Key Laboratory of Silicate Materials for Architectures, Wuhan University of Technology, Wuhan 430030, China; kaixu@whut.edu.cn
3 School of Chemical, Materials and Biological Engineering, The University of Sheffield, Sheffield S1 3JD, UK
* Correspondence: m.ojovan@sheffield.ac.uk

Polyphase materials—where crystalline and vitreous constituents coexist and interact—offer a uniquely rich design space in which functionality can be engineered through phase assemblage, nanoscale morphology, interfaces, and defect chemistry [1–7]. The present Special Issue, 'Synergy in Polyphase Materials: Harnessing the Power of Glass and Ceramics', assembles eight open access papers that collectively demonstrate how targeted processing strategies translate into controllable microstructures and, ultimately, measurable performance gains across optical, thermal, electrical, tribological, and bioactive domains. A central theme in this Special Issue is the shift from viewing multiphase systems as an unavoidable complexity to treating them as a deliberate platform for property tuning. This is particularly clear in glass–ceramics, where phase selection, crystal size distribution, and the chemistry of residual glass decisively determines both their functional responses and reliability. The presence of vitreous components in polyphase materials is crucial due to their universality, which is provided by the amorphous state of this solid phase of amorphous materials [8]. They enable unlimited variations of compositions tolerating arbitrary ratios between components of glass which can be defined as, following Michael Faraday, "a solution of different substances one in another". This is effectively used during the phase assemblage of composite materials, which is outlined in the contributions in this Special Issue, as well as in by many other dedicated topical collections such as [9,10]. Below, we briefly introduce the contributions to this Special Issue, highlighting their most important features.

Veselov et al. address a long-standing trade-off in transparent glass–ceramics: strengthening typically comes at the cost of optical scattering, while preserving transparency often limits mechanical performance [contribution 1]. By developing a sodium-modified $ZnO–MgO–Al_2O_3–SiO_2$ (ZMAS) glass that remains fully compatible with two-stage crystallization into exclusively $ZnAl_2O_4$ (gahnite) nanocrystals, the authors enable an additional strengthening degree of freedom—chemical ion exchange in molten KNO_3—without disrupting phase purity or optical transmittance. Importantly, the work underscores a broader, transferable concept: in transparent polyphase systems, mechanical robustness can be enhanced most effectively when bulk nanocrystallization and surface compressive-stress engineering are coupled rather than being treated as separate, potentially competing, interventions. This opens up new possibilities for electronic and photonic devices.

A complementary glass–ceramic contribution examines how redox conditions during melting influence phase transformations and defect-related optical responses in TiO_2-nucleated lithium aluminosilicate (LAS) glass–ceramics. Maltsev et al. show that varying melting conditions (including the use of As_2O_3) change the kinetics of phase evolution and the appearance of optically active reduced titanium species (Ti^{3+} and Ti^{3+}–Ti^{4+} pairs), while the overall transformation sequence remains largely preserved [contribution 2]. Beyond the specific LAS system, this study highlights an essential point for the design of polyphase materials: controlling redox and variable-valence species is not merely a "chemical detail", but rather a decisive lever that couples directly to nucleation dynamics, intermediate amorphous-region chemistry, and the eventual functional signature of multiphase products.

The Special Issue also presents an instructive example of how unconventional thermal processing can be exploited to enhance optical performance in ceramic pigment systems. Monrós et al. explore delafossite cuprates $CuMO_2$ (M = Mn, Fe, Cr) and doped/composite variants, aiming to approach ultra-black absorption metrics [contribution 3]. A key outcome is the demonstrated advantage of microwave-assisted firing in improving blackness compared to conventional electric kiln treatment, including cases where composite effects between the delafossite phase and associated chromium spinel become beneficial. The broader implication is methodological: for complex oxides where functional appearance sensitively depends on phase purity, defect populations, and microstructural uniformity, the heating modality itself can act as a "processing dopant", accelerating reactions and shifting the balance between competing formation pathways.

Energy-related multiphase engineering is represented by the thermoelectric study of Thiem et al. on $NbCoNi_xSn$ (x = 0–1), a system in which controlled introduction of Ni drives a transition from half-Heusler solid solutions at low x to half-Heusler/full-Heusler composites at higher x [contribution 4]. The work is valuable not only because it reports a composition window with an enhanced thermoelectric figure of merit (with the best performance around low Ni content), but also because it clarifies how interstitial occupation, secondary-phase emergence, and defect-driven phonon scattering co-evolve with electronic transport. In a field where incremental gains often depend on managing inherently multiphase microstructures, this paper contributes a coherent structure–property narrative that can guide rational optimization rather than purely empirical compositional screening.

The importance of enabling components—materials that do not constitute the primary functional phase, but determine real-world performance—appears in the tribology-focused contribution by Jianrong Liu et al., who developed a boron-modified phenolic resin for automotive friction materials [contribution 5]. By introducing B–O bonds and combining chemical modification with nano-alumina addition, the authors increased thermal decomposition resistance and report improvements in thermal fade behavior, supported by observations that are consistent with the formation of a more continuous friction film during braking. This study reinforces a practical message relevant to many polyphase systems: functional performance is frequently governed by interfacial integrity and the stability of transient, mechanically generated surface layers, making the chemistry of binders and secondary constituents as consequential as the selection of the "main" ceramic ingredients.

Two contributions address $BaTiO_3$-derived perovskite ceramics, emphasizing how processing control shapes dielectric response through microstructural evolution. Feliksik et al. investigate $Ba_{0.75}Ca_{0.25}TiO_3$ (BCT) ceramics and systematically connect the final sintering temperature to grain development, microstructural features, and dielectric behaviors [contribution 6]. Such processing–property coupling is central to perovskites because subtle shifts in density, grain-boundary characteristics, and defect distributions can dominate losses and stability, often more strongly than nominal composition alone. In a related but technologically distinct format, Zou et al. report Nb^{5+}-doped BCZT-based ceramic thick films prepared via a scraping process and examine how sintering temperature influences the resulting microstructure and functional response in film form [contribution 7]. Together, these two studies articulate a shared conclusion: whether in bulk or thick-film architectures,

https://doi.org/10.3390/ma19030478

reproducible dielectric/piezoelectric functionality requires process windows that are explicitly tuned to control microstructural heterogeneity rather than relying on composition as the sole "design variable".

Finally, the bioactivity-oriented work by Qidong Liu et al. expands the scope of this Special Issue toward amorphous ceramics designed for controlled reactivity [contribution 8]. By modifying silicon oxycarbide (SiOC) with Ca or Mg and evaluating network depolymerization indicators alongside biological assays and simulated body fluid immersion, the study demonstrates that changes in network connectivity correlate with apatite formation behavior, while Ca- and Mg-modified compositions exhibit distinct response kinetics. This contribution underscores an important conceptual parallel with the other papers: in polyphase (or structurally complex) materials, performance is often dictated by how the "non-primary" structural motifs—here, network modifiers and non-bridging oxygen populations—govern interfacial reactions and time-dependent functional outcomes.

In summary, this Special Issue highlights the breadth and depth of polyphase design strategies for glass and ceramics. Across the contributions, recurring mechanisms (controlled crystallization, redox-mediated defect chemistry, processing-route-enabled phase selection, and interface/stress engineering) reappear in different material families and application contexts. The resulting picture is not one of fragmented subfields, but of a shared methodology: harnessing phase coexistence and interfacial phenomena as purposeful tools for property optimization. We anticipate that the insights provided here will stimulate further cross-fertilization between glass science, functional ceramics, and applied materials engineering, and also that they will support the development of next-generation polyphase materials where performance is not merely observed, but designed.

Conflicts of Interest: The authors declare no conflicts of interest.

List of Contributions:

1. Veselov, I.; Shakhgildyan, G.; Savinkov, V.; Golubev, N.; Tregubov, K.; Vinogradov, D.; Avakyan, L.; Ojovan, M.; Ghosh, M.; Sigaev, V. Ultra-Strong Transparent ZnAl2O4 Glass-Ceramics via Controlled Crystallization and Ion Exchange. *Materials* **2025**, *18*, 5230. https://doi.org/10.3390/ma18225230.
2. Maltsev, S.; Dymshits, O.; Alekseeva, I.; Volokitina, A.; Tenevich, M.; Bachina, A.; Bogdanov, K.; Zapalova, S.; Shakhgildyan, G.; Zhilin, A. Glass-Ceramics of the Lithium Aluminosilicate System Nucleated by TiO_2: The Role of Redox Conditions of Glass Melting in Phase Transformations and Properties. *Materials* **2025**, *18*, 785. https://doi.org/10.3390/ma18040785.
3. Monrós, G.; Esteve, V.; Delgado, C.; Monrós-Andreu, G.; Llusar, M. In Search of Ultra-Black Ceramic Pigments Using Microwaves: Delafossite Cuprates $CuMO_2$ (M = Mn, Fe, Cr). *Materials* **2025**, *18*, 4910. https://doi.org/10.3390/ma18214910.
4. Thiem, M.; Yan, R.; Weidenkaff, A.; Xie, W. Thermoelectric Properties of $NbCoNi_xSn$ (x = 0 1). *Materials* **2025**, *18*, 3189. https://doi.org/10.3390/ma18133189.
5. Liu, J.; Guo, J.; Deng, J.; Fan, S.; Cai, X.; Kou, S.; Yang, S. Preparation and Properties of Boron Modified Phenolic Resin for Automotive Friction Materials. *Materials* **2025**, *18*, 1624. https://doi.org/10.3390/ma18071624.
6. Feliksik, K.; Adamczyk-Habrajska, M.; Makowska, J.; Bartkowska, J.A.; Pikula, T.; Panek, R.; Starczewska, O. The Impact of the Final Sintering Temperature on the Microstructure and Dielectric Properties of $Ba_{0.75}Ca_{0.25}TiO_3$ Perovskite Ceramics. *Materials* **2024**, *17*, 5210. https://doi.org/10.3390/ma17215210.
7. Zou, Y.; Fang, B.; Lu, X.; Zhang, S.; Ding, J. Preparation and Properties of Nb^{5+}-Doped BCZT-Based Ceramic Thick Films by Scraping Process. *Materials* **2024**, *17*, 4348. https://doi.org/10.3390/ma17174348.
8. Liu, Q.; Chen, H.; Wu, X.; Yan, J.; Yang, B.; Shi, C.; Li, Y.; Yu, S. Microstructure and Bioactivity of Ca- and Mg-Modified Silicon Oxycarbide-Based Amorphous Ceramics. *Materials* **2024**, *17*, 6159. https://doi.org/10.3390/ma17246159.

References

1. Höland, W.; Beall, G.H.; Smith, C.M. Glass-ceramics: From ideas to products. *J. Am. Ceram. Soc.* **2025**, *108*, e20086. [CrossRef]
2. Naumov, A.S.; Sigaev, V.N. Transparent Lithium-Aluminum-Silicate Glass-Ceramics (Overview). *Glass Ceram.* **2024**, *80*, 491–499. [CrossRef]
3. Yang, Y.; Tang, Y.; Jiang, G.; Fang, G.; Xue, Y.; Tan, Z.; Pan, D. Advances in glass-ceramics raw material selection-performance modulation-application expression: A review. *Mater. Today* **2025**, *89*, 270–292. [CrossRef]
4. McCloy, J.S.; Schuller, S. Vitrification of wastes: From unwanted to controlled crystallization, a review. *Compt. Rendus Geosci.* **2022**, *354*, 121–160. Available online: https://comptes-rendus.academie-sciences.fr/geoscience/articles/10.5802/crgeos.111/ (accessed on 6 January 2026). [CrossRef]
5. Zanotto, E.D. A bright future for glass-ceramics. *Am. Ceram. Soc. Bull.* **2010**, *89*, 19–27.
6. Rawlings, R.D.; Wu, J.P.; Boccaccini, A.R. Glass-ceramics: Their production from wastes—A Review. *J. Mater. Sci.* **2006**, *41*, 733–761. [CrossRef]
7. Colombo, P.; Brusatin, G.; Bernardo, V.; Scarinci, G. Inertization and reuse of waste materials by vitrification and fabrication of glass-based products. *Curr. Opin. Solid State Mater. Sci.* **2003**, *7*, 225–239. [CrossRef]
8. Ojovan, M.I.; Louzguine-Luzgin, D.V. Role of Structural Changes at Vitrification and Glass–Liquid Transition. *Materials* **2025**, *18*, 3886. [CrossRef] [PubMed]
9. Ojovan, M.I.; Darmaev, M.V.; Mashanov, A.A. Advances in glass and glass crystalline materials: An overview. *AIMS Mater. Sci.* **2025**, *12*, 1241–1245. [CrossRef]
10. Shakhgildyan, G.; Ojovan, M.I. *Advanced Glasses and Glass-Ceramics*; MDPI: Basel, Switzerland, 2024; 168p. [CrossRef]

https://doi.org/10.3390/ma19030478

materials

MDPI

Article

Ultra-Strong Transparent $ZnAl_2O_4$ Glass-Ceramics via Controlled Crystallization and Ion Exchange

Ivan Veselov [1], Georgiy Shakhgildyan [1,*], Vitaliy Savinkov [1], Nikita Golubev [1], Kirill Tregubov [1], Daniil Vinogradov [1], Leon Avakyan [2], Michael Ojovan [3], Manasi Ghosh [4] and Vladimir Sigaev [1]

1 Department of Glass and Glass-Ceramics, Mendeleev University of Chemical Technology, Moscow 125047, Russia; ivan.veselov.115@gmail.com (I.V.); sav.vi@mail.ru (V.S.); golubev_muctr@mail.ru (N.G.); kirill.tregubov.2018@gmail.com (K.T.); vdvhostagedaniil@yandex.ru (D.V.); vlad.sigaev@gmail.com (V.S.)
2 Physics Faculty, Southern Federal University, Rostov-on-Don 344090, Russia; laavakyan@sfedu.ru
3 School of Chemical, Materials and Biological Engineering, The University of Sheffield, Sheffield S1 3JD, UK; m.ojovan@sheffield.ac.uk
4 Physics Section, MMV, Banaras Hindu University, Varanasi 221005, India; manasi.ghosh@bhu.ac.in
* Correspondence: shakhgildian.g.i@muctr.ru

Abstract

Enhancing the mechanical strength of transparent glass-ceramics (TGCs) without compromising their optical performance remains a key challenge for advanced optical and photonic materials. Among aluminosilicate systems, ZnO–MgO–Al_2O_3–SiO_2 (ZMAS) glasses are particularly attractive due to their ability to form $ZnAl_2O_4$-based nanostructures; however, their ion-exchange (IE) strengthening has not been systematically explored due to the absence of single-charged cations in their composition. In this study, a sodium-modified ZMAS glass was developed to enable efficient chemical strengthening while preserving glass-forming ability and optical clarity. Controlled two-stage heat treatment produced TGCs containing 5 mol% Na_2O, composed solely of $ZnAl_2O_4$ (gahnite) nanocrystals with an average size of 4–5 nm. The obtained TGCs showed a Vickers hardness of ~8.5 GPa, increasing to ~10–10.5 GPa after ion exchange in molten KNO_3 at 450 °C, without changes in phase composition or optical transmittance. Compared with literature data on alkali-containing TGCs, the developed material demonstrates a higher hardness level while maintaining full transparency. The results reveal a practical route toward chemically strengthened $ZnAl_2O_4$-based glass-ceramics combining optical clarity, high hardness, and damage tolerance for optical, photonic, and protective applications.

Keywords: transparent glass-ceramics; glass crystallization; gahnite; ion exchange; Vickers hardness; glass strengthening

1. Introduction

Improving the mechanical reliability of transparent glasses and glass-ceramics remains a key challenge for applications in aerospace systems, optoelectronics, precision optics, and photonic devices, where high optical clarity must be combined with resistance to surface damage and cracking. Two industrially mature routes are widely used to increase strength: thermal tempering [1] and chemical strengthening by ion exchange (IE) [2–4]. In parallel, the transition from a homogeneous glass to a glass-ceramic, via controlled crystallization, often provides a substantial gain in hardness and damage tolerance while preserving transparency when the microstructure is nanometric and uniformly distributed [5–8].

Ion-exchange strengthening of transparent glass-ceramics (TGCs) is particularly attractive because a compressive surface layer can be superimposed on a fine, mechanically robust glass-ceramic framework. Successful IE of glass-ceramics, however, requires that the residual glassy phase contains a sufficient concentration of alkali cations amenable to substitution ($Li^+/Na^+ \rightarrow K^+$). Compositional tuning to increase the alkali content may, in turn, alter the crystallization pathway—phase selection, kinetics, and even the balance between bulk and surface crystallization—and thereby jeopardize transparency if coarse or mismatched phases precipitate.

Among aluminosilicate families, the most explored chemistries for TGCs strengthened by IE include Li_2O–Al_2O_3–SiO_2 (LAS) [9–12], MgO–Al_2O_3–SiO_2 (MAS) [13], ZnO–Al_2O_3–SiO_2 (ZAS) [14–17], and their combinations (e.g., ZMAS) [13,18,19]. To enhance IE efficiency, Na_2O is often introduced into LAS-type glasses to facilitate $Li^+ \rightarrow Na^+ \rightarrow K^+$ exchange, either in one step ($KNO_3/NaNO_3$ melts) or in staged protocols (Na^+ exchange followed by K^+ exchange) [10,15]. Table 1 compiles representative literature data for TGCs containing alkali species, covering salt compositions, IE schedules, and the resulting mechanical properties. Despite differences in test conditions (especially Vickers loads), these reports generally place pre-IE Vickers hardness in the range ≈6–8 GPa; post-IE strengthening typically raises microhardness to ≈7–9 GPa.

Table 1. Literature data on the composition and properties of TGC before and after ion-exchange strengthening.

Glass System and Alkali Oxide Content (mol.%)	Main Crystalline Phase	Salt Composition and IE Treatment Conditions	Vickers Hardness, GPa (Load, kgf)	Ref.
LAS Li_2O—21 Na_2O—1	$Li_2Si_2O_5$ $LiAlSi_4O_{10}$	Before IE $Na_{0.95}K_{0.05}NO_3$ 410 °C—4 h 410 °C—6 h 410 °C—8 h 410 °C—10 h	7.78 (n/a) 8.53 (n/a) 8.87 (n/a) 8.64 (n/a) 8.55 (n/a)	[9]
LAS Li_2O—15 Na_2O—9 K_2O—1	$LiAlSiO_4$	Before IE $(Na_{0.8}K_{0.2})NO_3$ 420 °C—4 h	7.03 (0.2) 7.62 (0.2)	[10]
LAS Li_2O—21.6	$LiAlSi_4O_{10}$	Before IE $NaNO_3$ 450 °C—6 h	6.95 (0.2) 7.45 (0.2)	[11]
LS Li_2O—27.5	$Li_2Si_2O_5$	Before IE $NaNO_3$ 315 °C—12 h 385 °C—1 h 450 °C—1 h	5.9 (0.05) 6.5 (0.05) 6.3 (0.05) 6.2 (0.05)	[12]
ZMAS Na_2O—8	$ZnAl_2O_4$ $MgAl_2O_4$	Before IE KNO_3 440 °C—4 h	7.2 (0.2) 8.3 (0.2)	[13]
ZAS Na_2O—8.5	Zn_2TiO_4 Zn_2SiO_4	Before IE KNO_3 410 °C—4 h 430 °C—4 h 450 °C—4 h	6.6 (0.2) 7.8 (0.2) 7.7 (0.2) 7.6 (0.2)	[14]

Table 1. *Cont.*

Glass System and Alkali Oxide Content (mol.%)	Main Crystalline Phase	Salt Composition and IE Treatment Conditions	Vickers Hardness, GPa (Load, kgf)	Ref.
LZAS Li_2O—5.95	$ZnAl_2O_4$ $LiAlSi_4O_{10}$	Before IE $NaNO_3$ 460 °C—4 h -> KNO_3 460 °C—4 h	7.45 (0.2) 8.1 (0.2)	[15]
ZAS Na_2O–10.5	ZnO Zn_2SiO_4	Before IE KNO_3 460 °C—8 h	6.3 (0.2) 7.6 (0.2)	[16]
LAS Li_2O—12.4 Na_2O—2.8	β-SiO_2 $LiAlSi_2O_6$ Li_2SiO_3	Before IE KNO_3 400 °C—10 h 420 °C—10 h 440 °C—10 h	5.2 (0.01) 7.3 (0.01) 7.4 (0.01) 7.7 (0.01)	[18]
NAS Na_2O—15	$NaAlSiO_4$	Before IE KNO_3 600 °C—8 h	4.4 (n/d) 6.3(n/d)	[20]
LAS Li_2O—15 Na_2O—9	$LiAlSiO_4$ $NaAlSiO_4$	Before IE $Na_{0.85}K_{0.15}NO_3$ 430 °C—4 h 450 °C—4 h 470 °C—4 h 490 °C—4 h -> KNO_3 380 °C—4 h 400 °C—4 h 420 °C—4 h 440 °C—4 h	7.4 (0.2) 7.5 (0.2) 7.5 (0.2) 7.5 (0.2) 7.5 (0.2) 8.0 (0.2) 8.1 (0.2) 8.2 (0.2) 8.1 (0.2)	[21]
ZMAS Na_2O—5.9	$MgAl_2O_4$ $ZnAl_2O_4$ SnO_2 ZrO_2	Before IE $Na_{0.02}K_{0.98}NO_3$ 450 °C—5 h 470 °C—5 h 490 °C—5 h	7.0 (0.2) 8.0 (0.2) 8.1 (0.2) 8.2 (0.2)	[22]

Within ZMAS-type systems, IE compatible formulations pose a particular challenge. For example, in a Na_2O-rich composition (6ZnO–12MgO–8Na_2O–17Al_2O_3–52.5SiO_2–3TiO_2–1.5ZrO_2) [13], part of Na_2O is consumed by Na_2TiSiO_5 crystallization alongside spinels ($ZnAl_2O_4$/$MgAl_2O_4$); the highest reported hardness (~8.2 GPa) was accompanied by a transmittance of only ~60–80% (380–780 nm) for 2 mm thick samples, likely due to coarse (>40 nm) crystals. In contrast, low-Na ZMAS glass (1.1 mol% Na_2O) crystallized into a $ZnAl_2O_4$-based TGC with microhardness near 9 GPa [23]; however, the low alkali content limits the subsequent IE efficiency. Recent work demonstrated that increasing Na_2O to 5 mol% preserves glass-forming ability (bulk casts ~200 g without devitrification) and enables moderate strengthening of the glass by IE (from 6.7 to 8.1 GPa) [24]. The crystallization behavior of this 5 mol% Na_2O ZMAS composition, its ability to yield a transparent spinel-based TGC, and its response to ion exchange have not yet been established.

Despite extensive studies on lithium- and magnesium-aluminosilicate systems, the development of chemically strengthened transparent glass-ceramics in the ZnO–MgO–Al_2O_3–SiO_2 family remains limited. The main challenge lies in achieving an alkali-rich residual glass phase—necessary for efficient ion exchange—while maintaining controlled bulk crystallization that preserves optical transparency.

In this context, the present study explores a Na_2O-modified ZMAS glass designed to balance glass-forming ability, nanocrystallization behavior, and ion-exchange capability. We show that moderate Na_2O incorporation enables the formation of a fully transparent $ZnAl_2O_4$-based nanostructure suitable for subsequent ion-exchange strengthening without altering the phase composition or optical clarity. This approach advances the understanding of how nanostructural design and ion-exchange strengthening can be effectively combined in ZMAS-type glass-ceramics, opening opportunities for the development of mechanically durable transparent materials for optical, photonic, and protective applications.

2. Materials and Methods

2.1. *Glass Composition and Batching*

The target (nominal) glass composition investigated in this work (hereafter 5Na, mol%) was: 20.1 ZnO; 10.0 MgO; 5.0 Na_2O; 9.8 Al_2O_3; 48.0 SiO_2; 5.9 TiO_2; 1.2 ZrO_2. This composition features a ZnO/MgO ratio of ~2:1 and thus differs markedly from the formulation reported in [13], where ZnO/MgO $\approx$ 1:2. Analytical-grade raw materials were used: SiO_2 ($\geq$99.99%), ZrO_2 ($\geq$99.99%), ZnO ($\geq$99.5%), TiO_2 ($\geq$99.99%), Na_2CO_3 ($\geq$99.8%), MgO ($\geq$98.0%), and $Al(OH)_3$ ($\geq$97.0%). The weighed powders were homogenized by mechanical mixing for 4 h and then charged into a corundum crucible.

2.2. *Melting, Casting, and Annealing*

Melting was carried out in a bottom-loading furnace equipped with $MoSi_2$ heating elements. The batch was loaded into a corundum crucible (Al_2O_3, 99.9%) and heated to 1550 °C, held for 1 h at the maximum temperature, and the melt was then cast into a preheated metal mold to produce bulk glass. The castings were annealed at 550 °C for 4 h in a muffle furnace to relieve residual internal stresses.

2.3. *Controlled Crystallization*

Annealed glasses were cut and polished, then subjected to a two-step heat treatment in a muffle furnace. The nucleation stage was performed at 660 °C for 4 h (in the vicinity of Tg), followed by a crystal-growth stage conducted at temperatures near the first DSC exotherm for 10–110 h. The resulting TGC are denoted 5Na–X–Y, where X is the heat treatment temperature and Y is the holding time (h).

2.4. *Ion Exchange Treatment*

Ion exchange (IE) of both the parent glass and the TGC was conducted in a pure KNO_3 melt at 450 °C for 12, 24, 48, 72, 96, and 120 h. After IE, samples were withdrawn from the salt bath and subjected to inertial cooling in a preheated muffle furnace to minimize thermal shock and relax thermal stresses. Residual salts were removed by rinsing in distilled water.

2.5. *Characterization of Structure and Properties*

Differential scanning calorimetry (DSC) was performed by means of a thermal analyzer STA 449 F3 Jupiter (Netzsch, Selb, Germany) in a Pt crucible, at a heating rate of 10 °C/min in Ar. The glass transition temperature (Tg) was determined as the extrapolated onset of the transition, while exopeak temperatures (Tp) were defined as the peak extremum temperature in DSC curves. X-ray diffraction (XRD) patterns of powdered samples were recorded by means of a diffractometer D2 Phaser (Bruker, Karlsruhe, Germany) employing nickel-filtered CuKα radiation. Crystalline phases were identified by comparing the peak position and relative intensities in the X-ray diffraction pattern with the ICDD PDF-2 database. XRD data were collected between 10 and 60 2θ, with a step size of 0.02° and a counting time of 0.3 s per step. XRD patterns were refined using the Rietveld method as implemented in the GSAS-II v5.6.0 code [25]. Optical transmission spectra were recorded over 300–800 nm using a UV-Vis spectrophotometer (UV-3600, Shimadzu, Kyoto, Japan) on the 1.5 mm thick samples. The microstructure of the glass and TGC samples was examined using high-resolution transmission electron microscopy (HRTEM, JEM-2100Plus, JEOL, Peabody, MA, USA) operated at an accelerating voltage of 200 kV. Bulk glass samples were ground to fine powders in an agate mortar and ultrasonically dispersed in ethanol. A drop of the resulting suspension was deposited onto a carbon-coated copper grid and dried under ambient conditions for approximately 20 min. Bright-field TEM images were recorded from multiple regions of each grid to obtain statistically representative microstructural data. The digital micrographs were analyzed using ImageJ software (version 1.53n). NTEGRA Spectra Spectrometer (NT-MDT Co., Zelenograd, Moscow, Russia) with the Ar laser beam (488 nm excitation wavelength) was used to record Raman spectra of the bulk samples.

Vickers microhardness was measured using an HVS-1000 tester (Laizhou Huayin Testing Instrument Co., Ltd., Laizhou, Shandong, China) under a 200 g load with a 10 s dwell time. The Vickers hardness was calculated as

$$H_V = 1.854\,\frac{F}{d^2} \quad (1)$$

where F is the applied load and d is the mean indentation diagonal. For each specimen, ≥20 indents were performed for statistical reliability.

3. Results and Discussion

3.1. Structure and Thermal Behavior of the Parent Glass

The glass composition containing 5 mol% Na_2O was selected based on our previous research [24]. An increase in Na_2O content beyond this level led to partial crystallization during casting, hindering the formation of a homogeneous bulk glass. To determine the characteristic thermal transitions, DSC was performed on monolithic samples of the 5Na glass.

As shown in Figure 1a, the DSC curve exhibits a distinct glass transition at 662 °C, followed by two exothermic peaks at 760 °C (Tp_1) and 960 °C (Tp_2), corresponding to the primary and secondary crystallization stages, respectively. These data were used to establish the temperature parameters for controlled nucleation and crystal growth in the subsequent heat-treatment procedure.

To evaluate the relationship between surface and bulk crystallization, DSC measurements were also conducted on glass samples of identical mass in both powdered and monolithic forms (Figure S1). The two DSC curves display nearly identical shapes, with only a slight shift of the exothermic peaks toward higher temperatures for the powdered sample. This minimal difference indicates that the crystallization of the 5Na glass proceeds predominantly via a bulk mechanism. This conclusion is further supported by the complete coincidence of the XRD patterns obtained for powdered and monolithic samples heat-treated under identical conditions (660 °C—4 h—760 °C—10 h, Figure S2).

The amorphous structure of the 5Na glass was confirmed by XRD and TEM analyses. The XRD pattern exhibits a single broad diffuse halo centered near $2\theta \approx 30°$ (Figure 1b), confirming the absence of crystalline phases. TEM images of the glass show overall homogeneous contrast, yet reveal faint nanoscale compositional modulations, indicative of liquid–liquid phase separation typical for ZMAS-type glasses (Figure 1c,d). The corresponding SAED patterns display diffuse rings without discrete reflections, confirming that both phases remain fully amorphous. Such phase separation is a well-known precursor for controlled crystallization in $ZnO/MgO–Al_2O_3–SiO_2$ glasses, providing chemically enriched regions that subsequently act as favorable sites for nucleation of crystallites during heat treatment.

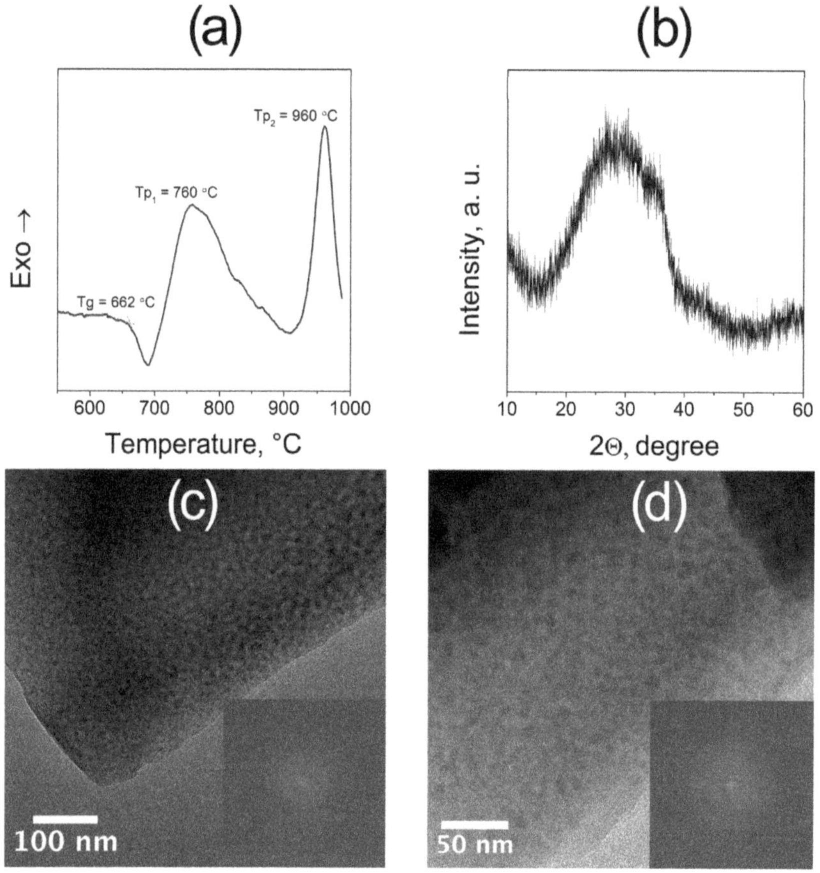

Figure 1. (**a**) DSC curve of the parent glass showing the glass transition temperature and two exothermic peaks corresponding to the sequential crystallization stages. (**b**) XRD pattern of parent glass confirming its amorphous structure. (**c**,**d**) HRTEM images of parent glass at different magnifications with corresponding SAED patterns (insets) showing diffuse rings typical of an amorphous matrix.

The relatively high glass transition temperature (Tg = 662 °C) of the 5Na composition, well above the typical ion-exchange temperature (≈450 °C), implies that structural relaxation during ion exchange is negligible, ensuring the stability of the glass matrix during chemical strengthening. Structural relaxation refers to the temperature-dependent rearrangement of the glass network toward equilibrium through viscous flow or atomic diffusion; at temperatures far below Tg, the glass structure remains rigid and cannot relax the stresses generated during ion exchange.

3.2. Crystallization Behavior and Phase Evolution

The temperature–time parameters for two-step crystallization were selected following the approach previously applied to transparent ZMAS glass-ceramics with 1.1 mol% Na_2O [23]. The nucleation stage was carried out at 660 °C for 4 h (near Tg), followed by a crystal growth stage within 750–850 °C for 10–110 h. Samples heat treated at 750–760 °C for 10 h retained full transparency, while increasing the temperature to 770 °C led to slight opalescence, and further heating to 785 °C and above produced opaque glass-ceramics.

The evolution of crystalline phases depending on temperature and holding time is shown in Figure 2. At 750–770 °C (Figure 2a), all diffraction peaks correspond to the zinc aluminate $ZnAl_2O_4$ (PDF #01-074-1138), indicating that a single spinel-type phase is formed in this temperature range. Further heating to 785–800 °C (Figure 2b) results in the appearance of additional reflections assigned to zinc silicate Zn_2SiO_4 (PDF #01-085-0453). The intensity of these peaks increases with temperature, and at 850 °C, weak peaks of forsterite Mg_2SiO_4 (PDF #01-076-0851) also appear. At this stage, silicate phases dominate over the spinel, causing a pronounced decrease in transparency.

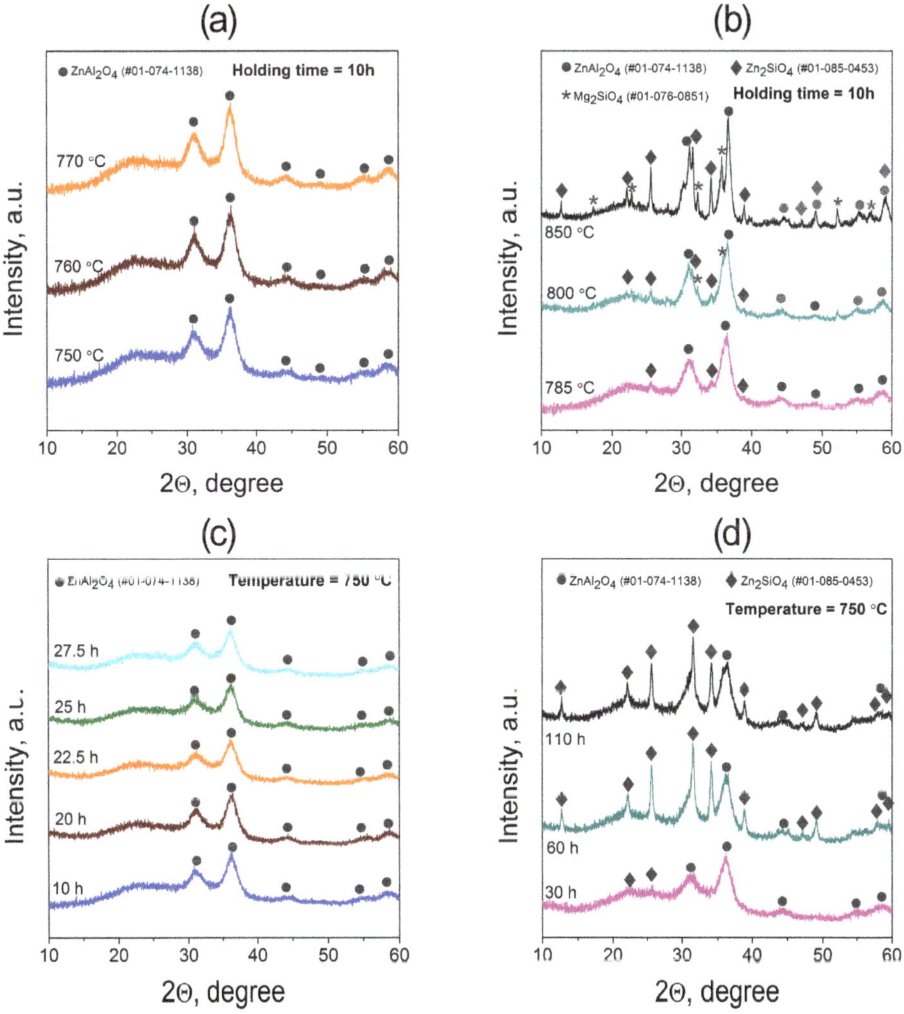

Figure 2. XRD patterns of the 5Na TGC heat-treated at different temperatures and durations: (**a**,**c**) transparent samples containing only $ZnAl_2O_4$ spinel; (**b**,**d**) opaque samples containing additional Zn_2SiO_4 and Mg_2SiO_4 phases.

For samples treated isothermally at 750 °C for 10–110 h (Figure 2c,d), the diffraction data reveal a time-dependent phase evolution. Up to ~25 h, $ZnAl_2O_4$ remains the only crystalline phase, while its mean crystallite size, calculated using the Scherrer equation from the (311) reflection at 36.3°, remains constant at ~4–5 nm. Prolonged holding (>60 h) promotes the formation of Zn_2SiO_4, accompanied by optical opacity due to increased scattering on larger (~25–60 nm) silicate crystallites.

A detailed Rietveld refinement was performed on the set of selected samples (5Na-750-10, 5Na-760-10, 5Na-850-10, and others). The refinement employed structural models from the Crystallography Open Database (COD) [26] for $ZnAl_2O_4$ (ID 9007017, 9 reflections), Zn_2SiO_4 (ID 9014832, 98 reflections), and Mg_2SiO_4 (ID 9000535, 46 reflections). The lattice parameters (a), coherent scattering domain sizes (D), and metal site occupancies (Zn:Mg ratios) were refined simultaneously. Partial cation substitution between Zn^{2+} and Mg^{2+} sites was allowed, resulting in simulated intensities closely matching the experimental profiles. The flexible background "Chebyschev-1" function with 10 parameters was used to subtract diffuse peaks of glass phases.

The experimental and calculated XRD patterns, along with corresponding fits, are shown in Figure 3, and the numerical results are summarized in Table 2. The refinement quality, expressed as the goodness of fit (GOF), ranged between 1.1 and 1.5, which is acceptable for nanocrystalline glass-ceramic systems with residual amorphous background. Representative fits for other samples are provided in the Supporting Information (Figures S3–S8).

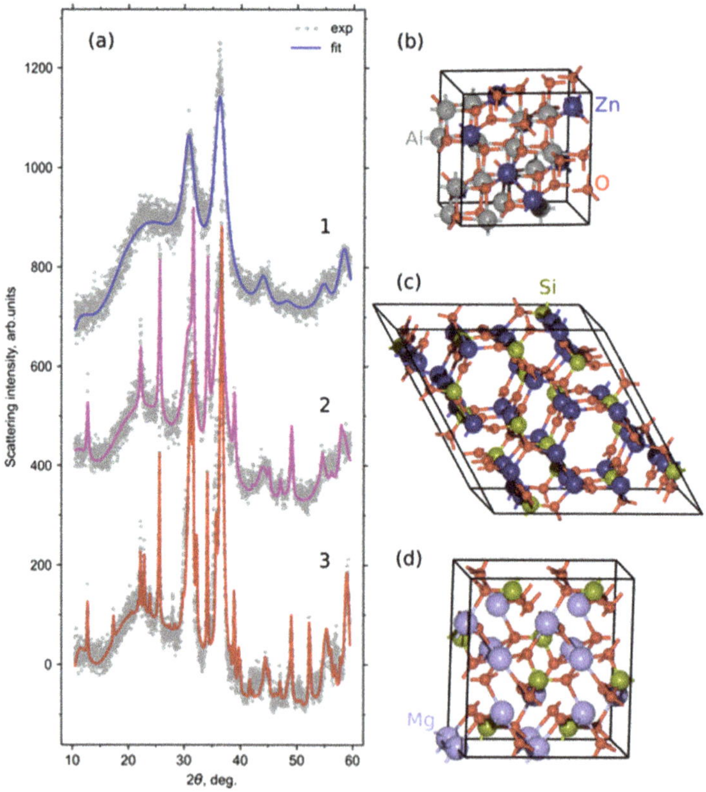

Figure 3. (**a**) Experimental XRD patterns (dots) and fitted theoretical curves (lines) obtained from Rietveld refinement for selected 5Na TGC: 1-5Na-750-10, 2-5Na-750-110, 3-5Na-850-10. Panels (**b–d**) illustrate the refined crystal structures of $ZnAl_2O_4$, Zn_2SiO_4, and Mg_2SiO_4 phases.

Table 2. Refined structural parameters of crystalline phases in 5Na TGC obtained from Rietveld analysis: lattice parameters (a), crystallite size (D), cation occupancies (Zn:Mg), phase fractions, and goodness of fit (GOF). The measurement uncertainty falling at the last digit is shown in brackets.

Sample	Crystalline Phase	Phase Fraction, wt.%	Cell Parameters, Å	Crystallite Size, nm	Site Occupancy Zn:Mg	GOF
5Na-750-10	$ZnAl_2O_4$	100	8.223 (3)	3.7 (1)	55:45 (2)	1.52
5Na-760-10	$ZnAl_2O_4$	100	8.207 (3)	4.1 (1)	52:48 (2)	1.27
5Na-750-25	$ZnAl_2O_4$	100	8.234 (3)	3.8 (1)	48:52 (2)	1.27
5Na-750-110	$ZnAl_2O_4$	83	8.220 (3)	3.9 (1)	57:43 (2)	1.35
	Zn_2SiO_4	17	13.905 (4), 9.302 (2)	25.8 (7)	45:55 (5)	
5Na-800-10	$ZnAl_2O_4$	86	8.205 (2)	5.0 (1)	62:38 (1)	1.39
	Zn_2SiO_4	5	13.94 (2), 9.28 (1)	12 (1)	100:0 *	
	Mg_2SiO_4	9	4.762 (4), 10.222 (7), 6.000 (5)	28.8 (4)	0:100 *	
5Na-850-10	$ZnAl_2O_4$	72	8.141 (2)	9.9 (2)	81:19 (3)	2.07
	Zn_2SiO_4	10	13.909 (4), 9.307 (2)	66 (5)	80:20 (10)	
	Mg_2SiO_4	18	4.769 (2), 10.239 (4), 5.982 (3)	56 (6)	15:85 (3)	
5Na-700-24	$ZnAl_2O_4$	80	8.204 (6)	3.3 (1)	49:51 (3)	1.12
	Zn_2SiO_4	7	14.51 (3), 8.38(1)	3 (1)	100:0 *	
	Mg_2SiO_4	13	4.69 (3), 10.09 (6), 6.08 (4)	7 (2)	0:100 *	
Reference	$ZnAl_2O_4$	-	8.085	-	-	-
	Zn_2SiO_4	-	13.971, 9.334	-	-	
	Mg_2SiO_4	-	4.752, 10.193, 5.977	-	-	

* Site occupancies for minor phases were fixed due to their small fraction.

A further improvement of the fitting quality may require variation in thermal vibration parameters, stresses, and size distributions influencing the XRD pattern. However, for the considered experimental data, such fine-tuning will give high cross-correlations between varied parameters and decrease the reliability of the obtained data. Also note that the interpretation of the obtained Zn:Mg ratios may be ambiguous since we cannot reliably distinguish the replacement of a heavier Zn atom (Z = 30) by a lighter Mg atom (Z = 12) and the formation of Zn vacancies by studying XRD data only. However, our interpretation is justified by the increase in electronic density in Mg silicate phases, which cannot be due to vacancy formations.

The refined parameters indicate that for transparent samples (e.g., 5Na-750-10 and 5Na-760-10), $ZnAl_2O_4$ is the only crystalline phase present. Its cubic lattice parameter is a = 8.21–8.23 Å, which is approximately 1% larger than that of the stoichiometric reference $ZnAl_2O_4$ (8.09 Å). This expansion likely results from partial substitution of Zn^{2+} by Mg^{2+} in tetrahedral sites, consistent with the refined cation occupancy ratios of Zn:Mg = 55:45–48:52.

With increasing temperature and holding time, the relative fraction of $ZnAl_2O_4$ decreases while silicate phases emerge. At 800 °C, the glass-ceramics consist of approximately 86 wt.% $ZnAl_2O_4$, 5 wt.% Zn_2SiO_4, and 9 wt.% Mg_2SiO_4, whereas at 850 °C, the phase composition shifts to 72 wt.% $ZnAl_2O_4$, 10 wt.% Zn_2SiO_4, and 18 wt.% Mg_2SiO_4. The silicate crystallites exhibit sizes between 25 and 60 nm, while spinel nanocrystals remain around 4–10 nm, preserving a fine-grained structure even at high temperatures.

No crystalline phases of ZrO_2 or TiO_2 were detected, which can be attributed to their low content in the base glass and the overlap of their most intense reflections with the broad spinel peaks. The refined results thus confirm that controlled crystallization at 750–760 °C for 10–25 h yields a uniform nanocrystalline $ZnAl_2O_4$ spinel phase with minor cation disorder (Zn–Mg intermixing), responsible for the combination of high transparency and mechanical stability.

Raman spectra of the parent glass and the glass-ceramics heat-treated at different temperatures are shown in Figure 4. The spectrum of the parent 5Na glass is broad and diffuse, exhibiting a weak, poorly resolved band near 800–850 cm^{-1}, which can be attributed to Si–O and Al–O stretching vibrations in a chemically heterogeneous glass network. Such spectral broadening is typical of glasses undergoing liquid–liquid phase separation, which is also evident from the TEM observations showing nanometric contrast variations in the glass matrix.

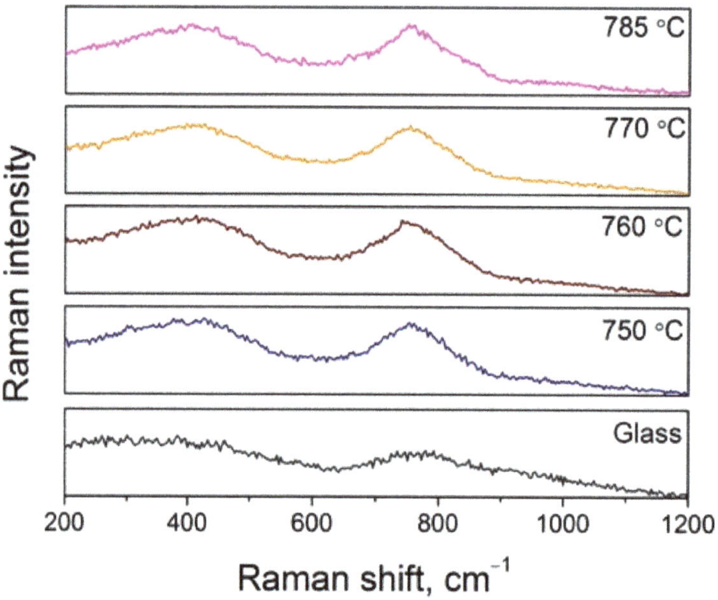

Figure 4. Raman spectra of the parent 5Na glass and TGCs heat-treated at different temperatures for 10 h.

Upon heat treatment at 750–785 °C, two distinct Raman bands become apparent at approximately 415 cm^{-1} and 760 cm^{-1}. The former corresponds to Zn–O symmetric stretching vibrations in the cubic spinel $ZnAl_2O_4$ [27,28], while the latter is assigned to Al–O symmetric stretching vibrations of $[AlO]_x$ units within the $ZnAl_2O_4$ lattice [29,30]. All spectra in this temperature range display the same set of features, and only the intensity of both bands increases slightly with temperature, indicating the growth and ordering of the gahnite nanocrystals rather than the formation of new phases. These observations confirm that the crystallization of the 5Na glass proceeds through phase-separation and the development of $ZnAl_2O_4$ nanocrystals, in full agreement with the XRD data.

Representative HRTEM images of the glass-ceramics heat-treated at 750 °C and 785 °C are shown in Figure 5. After heat treatment at 750 °C (Figure 5a), the material exhibits a homogeneous nanostructure with uniformly dispersed crystalline domains of $ZnAl_2O_4$ embedded in the glassy matrix. The nanocrystals are nearly spherical and their average size does not exceed 4–5 nm, consistent with the value estimated from XRD data. Lattice fringes corresponding to the (311) and (400) planes of cubic $ZnAl_2O_4$ are locally resolved, confirming the single-phase spinel structure of the crystalline component.

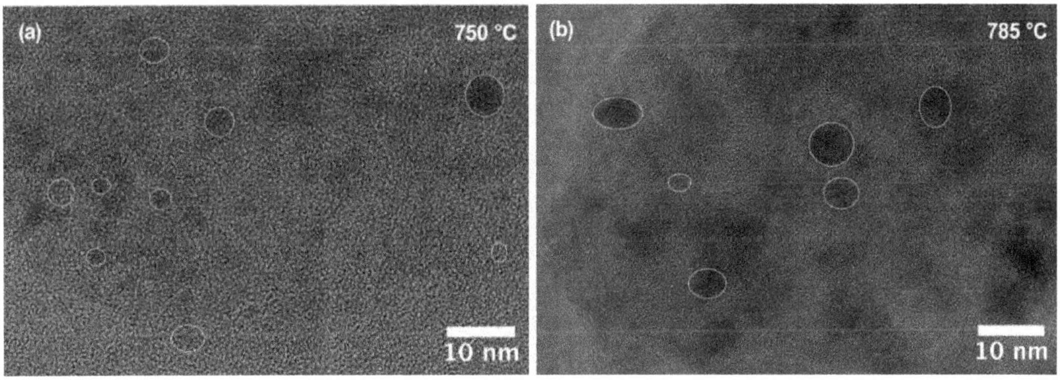

Figure 5. HRTEM images of TGC heat-treated at (**a**) 750 °C and (**b**) 785 °C. Yellow circles indicate $ZnAl_2O_4$ nanocrystals in the glassy matrix.

When the crystallization temperature is increased to 785 °C (Figure 5b), the contrast variations become more pronounced, and individual nanocrystals grow slightly in size, reaching 6–8 nm. Nevertheless, the material still maintains a fine and uniformly distributed nanocrystalline structure without the formation of secondary crystalline phases or large aggregates. The observed microstructural evolution supports the Raman and XRD results, indicating that the heat treatment up to 785 °C promotes growth and ordering of gahnite nanocrystals.

3.3. *Mechanical Properties and Ion-Exchange Strengthening*

The dependence of the Vickers hardness on the temperature of the crystal-growth stage is presented in Figure 6a. The parent 5Na glass shows an average hardness of 6.7 GPa. Two-stage heat treatment leads to a noticeable increase in hardness by about 1–2 GPa, yielding values in the range of 7.9–8.9 GPa depending on the temperature of the second stage. Samples treated at 750–760 °C remain fully transparent, while heating to 770 °C results in opalescence, and at 785 °C and above, the material becomes opaque. The highest hardness among the transparent glass-ceramics is achieved for the sample crystallized at 760 °C, reaching ≈8.5 GPa. The gradual rise in hardness up to 770–785 °C correlates with the growth and structural ordering of $ZnAl_2O_4$ nanocrystals revealed by TEM and Raman spectroscopy, whereas the subsequent decrease in hardness at higher temperatures is associated with the coarsening of crystallites and the appearance of secondary silicate phases, which deteriorate both the mechanical and optical properties.

The influence of IE in molten KNO_3 at 450 °C on the hardness of the parent glass and of the TGC 5Na–760–10 is shown in Figure 6b. For both materials, the hardness increases steadily with the duration of IE, indicating the formation of a compressive stress layer as Na^+ ions are replaced by K^+ in the surface region. During the first 24 h, the increase in hardness is more pronounced for the glass, likely due to the absence of crystal–matrix interfaces that hinder alkali diffusion. At longer treatment times ($\geq$ 24 h), the rate of hardness growth becomes similar for both materials. After 120 h, the hardness of the ion-exchanged TGC reaches ≈10–10.5 GPa, clearly exceeding that of the ion-exchanged glass. The higher strengthening efficiency in TGC arises from the restricted Na^+ mobility in the nanocrystalline spinel matrix, which promotes a steeper concentration gradient and higher compressive stress near the surface. Although saturation is not yet reached after 120 h, further improvement would likely be marginal and of limited technological relevance given the long processing time required.

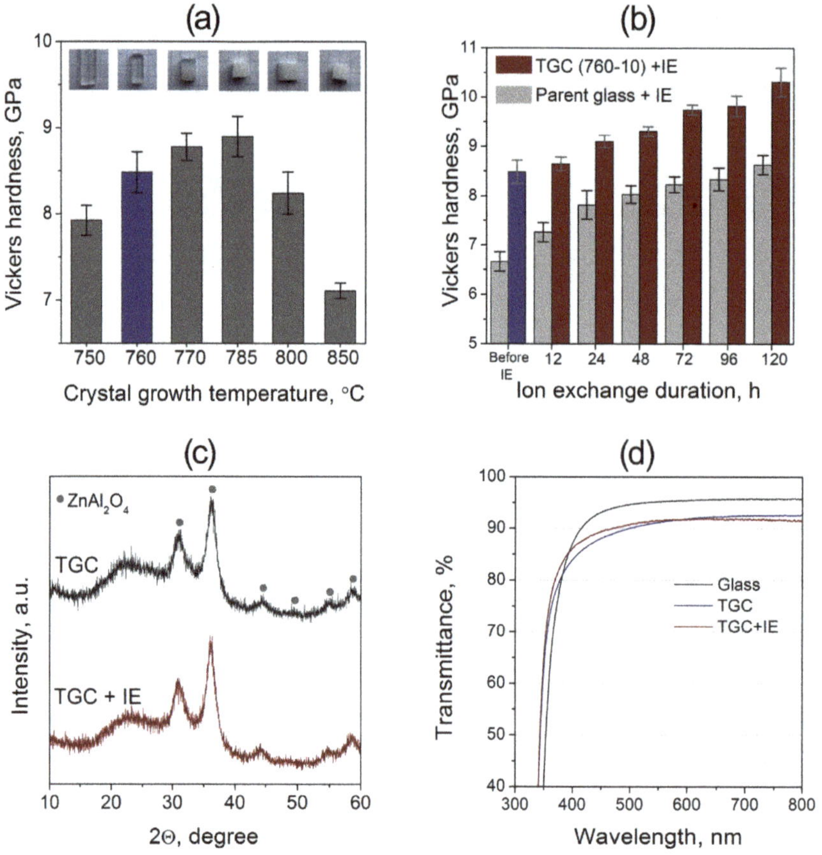

Figure 6. (**a**) Dependence of the Vickers hardness of 5Na glass-ceramics on the temperature of the crystal-growth stage (10 h). Insets show photographs of the samples illustrating the gradual loss of transparency with increasing crystallization temperature. (**b**) Evolution of Vickers hardness during IE for the parent glass and the TGC 5Na–760–10. (**c**) XRD patterns of the TGC before and after 120 h of IE. (**d**) Optical transmission spectra of the parent glass, the TGC, and the IE TGC.

To determine whether prolonged contact with molten salt affects the crystalline structure of the TGC, XRD patterns of the 5Na–760–10 sample were recorded before and after 120 h of IE (Figure 6c). The diffraction profiles are identical and fully indexed to $ZnAl_2O_4$, showing no additional peaks or shifts, which indicates that the IE process does not modify the phase composition or lattice parameters of the crystalline phase. The optical transmission spectra (Figure 6d) confirm that the ion-exchanged TGC retains high transparency in the 400–800 nm range, with only a minor reduction compared to the parent glass.

The optical micrographs of Vickers indents obtained under different loads for the parent glass, transparent glass-ceramic, and ion-exchanged glass-ceramic are shown in Figure 7. All imprints exhibit the characteristic square shape of the Vickers indenter. For the parent glass, radial cracks emanate from the corners of the indent even at the lowest applied load (0.2 kgf), and their number and length increase progressively with increasing load. After crystallization at 760 °C, the resulting TGC demonstrates a markedly improved crack resistance: no visible cracks are observed up to 0.5 kgf, and only at 1 kgf do minor damage zones appear around the indentation, associated with the brittle response of the residual glassy phase.

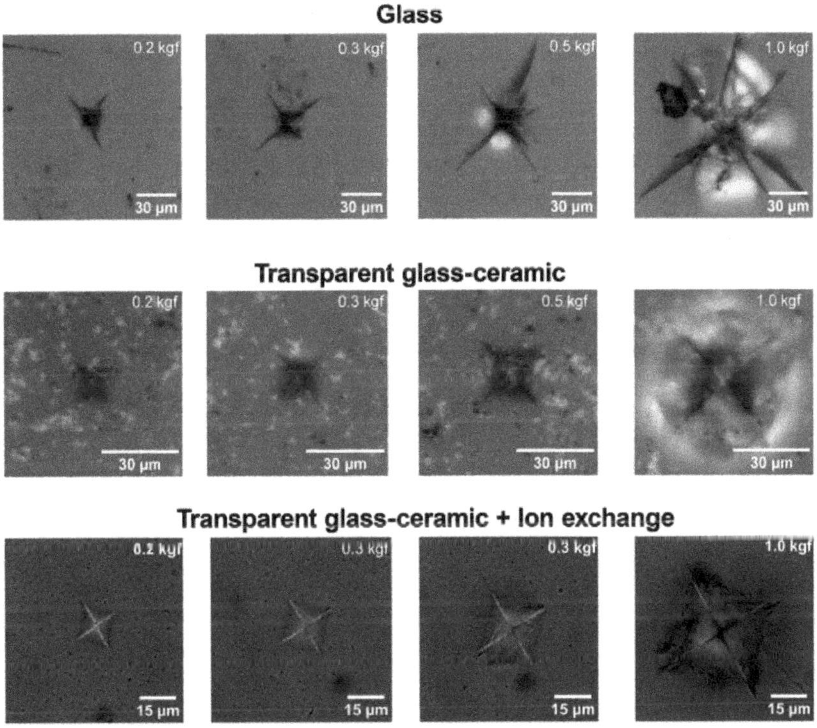

Figure 7. Optical micrographs of Vickers indents obtained under different loads (0.2–1.0 kgf) for (**top**) the parent glass, (**middle**) the TGC, and (**bottom**) the ion-exchanged TGC.

In contrast, the ion-exchanged TGC exhibits significantly enhanced resistance to indentation-induced fracture. Even at the maximum applied load (1 kgf), only small, short cracks appear from some indent corners, while the surrounding surface remains intact and free from mirror-like reflections typical of brittle fracture. This behavior confirms that the combination of a fine-grained $ZnAl_2O_4$ nanocrystalline framework and the compressive surface layer formed during ion exchange effectively suppresses crack propagation and improves the overall damage tolerance of the material.

Together with the hardness data, these results demonstrate that moderate crystallization at 750–760 °C followed by ion-exchange treatment in molten KNO_3 yields transparent glass-ceramics possessing both high surface hardness (>10 GPa) and outstanding resistance to cracking—properties crucial for mechanically loaded transparent components.

The obtained results can be compared with previously reported data summarized in Table 1. For most alkali-containing transparent glass-ceramics, the microhardness before ion exchange typically lies in the range of 6–8 GPa, increasing to 7–9 GPa after IE treatment [9–16,18,20]. In contrast, the present ZnO–MgO–Al_2O_3–SiO_2 system demonstrates higher hardness levels (8.5 GPa before IE and ≈10–10.5 GPa after IE) while maintaining high optical transparency in the 400–800 nm range. This performance surpasses that of previously studied ZMAS-type glass-ceramics [13,14], where coarse or mixed-phase crystallization limited transparency and further ion-exchange efficiency. The superior mechanical response of the present material can be attributed to the formation of a uniform nanocrystalline $ZnAl_2O_4$ spinel network combined with a compressive surface layer produced during ion exchange, which together ensure both high hardness and resistance to crack propagation.

4. Conclusions

Transparent ZMAS glass-ceramics containing 5 mol% Na_2O were successfully fabricated via a controlled two-stage heat treatment. The parent glass exhibited an amorphous structure with nanoscale compositional fluctuations, indicating the occurrence of liquid–liquid phase separation. DSC and XRD analyses revealed two exothermic crystallization events, while controlled crystallization at 660 °C for 4 h and 750–760 °C for 10 h produced transparent glass-ceramics consisting solely of $ZnAl_2O_4$ (gahnite) nanocrystals with an average size of 4–5 nm. Raman spectroscopy and HRTEM confirmed that increasing the temperature to 785 °C leads only to the growth and ordering of the spinel phase, without the formation of secondary silicates.

The optimal combination of high transparency and mechanical strength was achieved for samples crystallized at 760 °C, exhibiting a Vickers hardness of ≈8.5 GPa. Subsequent ion exchange in molten KNO_3 at 450 °C further increased the hardness to ≈10–10.5 GPa without altering the phase composition or transparency. The strengthened samples demonstrated a pronounced resistance to crack initiation and propagation under indentation.

Compared with literature data on alkali-containing transparent glass-ceramics (Table 1), the obtained material shows a higher hardness level while maintaining optical transparency, indicating the efficiency of the selected composition and processing route. Overall, the study demonstrates that moderate crystallization followed by ion exchange provides an effective route to produce transparent, mechanically durable $ZnAl_2O_4$-based glass-ceramics, combining optical clarity with enhanced surface hardness. Such materials are promising for applications requiring high optical transparency and mechanical reliability, including protective and structural components in optical and photonic devices, display and cover glasses, micro-optical elements, and transparent armor systems, where improved resistance to surface damage and cracking is critical. Future work will focus on evaluating the thermal-shock resistance, long-term stability, and scalability of the synthesis and ion-exchange strengthening processes to assess their potential for practical implementation.

Supplementary Materials: The following supporting information can be downloaded at: https://www.mdpi.com/article/10.3390/ma18225230/s1, Figure S1: DSC measurements on glass samples of identical mass in powdered and monolithic forms; Figure S2: XRD patterns for powdered and monolithic samples heat-treated under identical conditions 660 °C—4 h—760 °C—10 h; Figures S3–S8: The illustration of fitting qualities of the XRD patterns for the samples.

Author Contributions: Conceptualization, I.V. and G.S.; methodology, V.S. (Vitaliy Savinkov); validation, M.O. and M.G.; investigation, N.G., K.T. and D.V.; writing—original draft preparation, I.V. and G.S.; writing—review and editing, L.A. and V.S. (Vladimir Sigaev); visualization, I.V. and L.A.;

project administration, V.S. (Vladimir Sigaev); funding acquisition, V.S. (Vladimir Sigaev). All authors have read and agreed to the published version of the manuscript.

Funding: This work was supported by a Grant from the Russian Science Foundation (RSF Project No. 25-43-01075).

Institutional Review Board Statement: Not applicable.

Informed Consent Statement: Not applicable.

Data Availability Statement: The original contributions presented in this study are included in the article/Supplementary Material. Further inquiries can be directed to the corresponding author.

Conflicts of Interest: The authors declare no conflicts of interest.

References

1. Yang, P.; Wang, Y.; Li, X.; Han, T.; Shi, X. Effect of Physical Tempering on the Mechanical Properties of Aluminosilicate Glass. *J. Non Cryst. Solids* **2023**, *619*, 122574. [CrossRef]
2. Berneschi, S.; Righini, G.C.; Pelli, S. Towards a Glass New World: The Role of Ion-Exchange in Modern Technology. *Appl. Sci.* **2021**, *11*, 4610. [CrossRef]
3. Zhen, W.; Zhenbiao, H.; Tao, S.; Fenghua, Z.; Yulong, L.; Sheikh, M.Z.; Xiang, W.; Yinmao, W. A Comparative Study on the Effect of Loading Speed and Surface Scratches on the Flexural Strength of Aluminosilicate Glass: Annealed vs. Chemically Strengthened. *Ceram. Int.* **2018**, *44*, 11239–11256. [CrossRef]
4. Banapour Ghaffari, O.; Eftekhari Yekta, B.; Zakeri-Nasrabadi, M. Estimating "Depth of Layer" (DOL) in Ion-Exchanged Glasses Using Explainable Machine Learning. *Materialia* **2024**, *33*, 102027. [CrossRef]
5. Seidel, S.; Dittmer, M.; Wisniewski, W.; Höland, W.; Rüssel, C. Effect of the ZrO_2 Concentration on the Crystallization Behavior and the Mechanical Properties of High-Strength MgO–Al_2O_3–SiO_2 Glass–Ceramics. *J. Mater. Sci.* **2017**, *52*, 1955–1968. [CrossRef]
6. Lu, J.; Wang, H.; Li, Y.; Zhou, Y.; Jiang, W. Effect of Metastable Phase on the Crystallization and Mechanical Properties of MgO-Al2O3-SiO2 Glass-Ceramics without Nucleating Agents. *Ceram. Int.* **2023**, *49*, 7737–7745. [CrossRef]
7. Beall, G.H.; Comte, M.; Dejneka, M.J.; Marques, P.; Pradeau, P.; Smith, C. Ion-Exchange in Glass-Ceramics. *Front. Mater.* **2016**, *3*, 41. [CrossRef]
8. Shakhgildyan, G.Y.; Ojovan, M.I. Inhomogeneities in Glass: From Defects to Functional Nanostructures. *Encyclopedia* **2025**, *5*, 136. [CrossRef]
9. Zhang, T.; Zhang, Z.; Han, J.; Wang, J.; Liu, C.; Ruan, J. The Structure and Properties of Chemical Strengthened Transparent Lithium Disilicate Glass Ceramics with Various P_2O_5 Contents. *J. Non Cryst. Solids* **2022**, *588*, 121626. [CrossRef]
10. Mu, Y.; Qiu, J.; Wu, L.; Tian, X.; Wang, J.; Li, L.; Han, J. Ion-Exchange Induced Mechanical Properties Enhancement and Microstructural Modifications of Transparent Lithium Aluminosilicate Glass-Ceramics. *Ceram. Int.* **2024**, *50*, 14878–14883. [CrossRef]
11. Zheng, W.; Gao, Z.; Huang, M.; Zhang, H.; Yuan, J.; Tian, P.; Peng, Z.; Du, X. Chemical Strengthening of Lithium Aluminosilicate Glass-Ceramic with Different Crystallinity. *J. Non Cryst. Solids* **2022**, *598*, 121940. [CrossRef]
12. Li, X.C.; Meng, M.; Li, D.; Wei, R.; He, L.; Zhang, S.F. Strengthening and Toughening of a Multi-Component Lithium Disilicate Glass-Ceramic by Ion-Exchange. *J. Eur. Ceram. Soc.* **2020**, *40*, 4635–4646. [CrossRef]
13. Li, W.; Xiang, W.; Li, C.; Wang, J.; Li, L.; Han, J. Influence of Pre-Heating Treatment on the Crystallization and Ion-Exchange Behavior of MgO-Al2O3-SiO2 Glass-Ceramics. *Ceram. Int.* **2024**, *50*, 42933–42938. [CrossRef]
14. Wang, S.; Guo, Y.; Lu, Y.; Liu, C.; Deng, Z.; Han, N.; Cui, J. Microstructure and Ion-Exchange Properties of Transparent Glass-Ceramics Containing Zn_2TiO_4/α-Zn_2SiO_4 Nanocrystals. *J. Eur. Ceram. Soc.* **2022**, *42*, 3595–3602. [CrossRef]
15. Guo, Y.; Wang, J.; Ruan, J.; Han, J.; Xie, J.; Liu, C. Microstructure and Ion-Exchange Properties of Glass-Ceramics Containing $ZnAl_2O_4$ and β-Quartz Solid Solution Nanocrystals. *J. Eur. Ceram. Soc.* **2021**, *41*, 5331–5340. [CrossRef]
16. Guo, Y.; Liu, C.; Wang, J.; Ruan, J.; Xie, J.; Han, J.; Deng, Z.; Zhao, X. Effects of Alkali Oxides and Ion-Exchange on the Structure of Zinc-Alumino-Silicate Glasses and Glass-Ceramics. *J. Eur. Ceram. Soc.* **2022**, *42*, 576–588. [CrossRef]
17. Shakhgildyan, G.; Avakyan, L.; Atroshchenko, G.; Vetchinnikov, M.; Zolikova, A.; Ignat'eva, E.; Ziyatdinova, M.; Subcheva, E.; Bugaev, L.; Sigaev, V. Ultra-Broadband Plasmon Resonance in Gold Nanoparticles Precipitated in ZnO-Al_2O_3-SiO_2 Glass. *Ceramics* **2024**, *7*, 562–578. [CrossRef]
18. Lu, J.; Wang, H.; Zhu, J.; Zheng, Q.; Ding, L.; Jiang, W. Preparation and Characterization of High-Strength Glass-Ceramics via Ion-Exchange Method. *Materials* **2021**, *14*, 5477. [CrossRef]
19. Shakhgildyan, G.Y.; Alekseev, R.O.; Golubev, N.V.; Savinkov, V.I.; Naumov, A.S.; Presnyakova, N.N.; Sigaev, V.N. One-Step Crystallization of Gahnite Glass-Ceramics in a Wide Thermal Gradient. *ChemEngineering* **2023**, *7*, 37. [CrossRef]

20. Alzahrani, A.S. Novel Ion-Exchangeable Nepheline Glass-Ceramics for Dental Application. *J. Eur. Ceram. Soc.* **2023**, *43*, 5682–5690. [CrossRef]
21. Tian, X.; Chen, Z.; Mu, Y.; Ji, C.; Li, L.; Wu, L.; Qiu, J.; Wang, J.; Han, J. Promotion of Phase Transformation on Ion-Exchange: Transparent Glass-Ceramics with Ultra-High Flexural Strength and Impact Resistance. *J. Eur. Ceram. Soc.* **2026**, *46*, 117885. [CrossRef]
22. Xu, G.; Zhang, J.; He, H.; Dong, J.; Liu, J.; Wang, F.; Liao, Q.; Li, M.; Gu, Y.; Liu, L.; et al. The Effect of Ion Exchange on the Properties of MgO-Al_2O_3-SiO_2 Transparent Glass-Ceramics. *Ceram. Int.* **2025**, *51*, 33245–33253. [CrossRef]
23. Shakhgil'dyan, G.Y.; Savinkov, V.I.; Shakhgil'dyan, A.Y.; Alekseev, R.O.; Naumov, A.S.; Lopatkina, E.V.; Sigaev, V.N. Effect of Sitallization Conditions on the Hardness of Transparent Sitalls in the System ZnO–MgO–Al_2O_3–SiO_2. *Glass Ceram.* **2021**, *77*, 426–428. [CrossRef]
24. Veselov, I.A.; Naumov, A.S.; Savinkov, V.I.; Fedotov, S.S.; Alekseev, R.O.; Ignatieva, E.S.; Shakhgildyan, G.Y.; Sigaev, V.N. Ion Exchange Strengthening of Glasses of the ZnO–MgO–Al_2O_3–SiO_2 Glass-Ceramic-Forming System with an Increased Content of Na_2O. *Glass Ceram.* **2024**, *81*, 11–16. [CrossRef]
25. Toby, B.H.; Von Dreele, R.B. *GSAS-II*: The Genesis of a Modern Open-Source All Purpose Crystallography Software Package. *J. Appl. Crystallogr.* **2013**, *46*, 544–549. [CrossRef]
26. Gražulis, S.; Daškevič, A.; Merkys, A.; Chateigner, D.; Lutterotti, L.; Quirós, M.; Serebryanaya, N.R.; Moeck, P.; Downs, R.T.; Le Bail, A. Crystallography Open Database (COD): An Open-Access Collection of Crystal Structures and Platform for World-Wide Collaboration. *Nucleic Acids Res.* **2012**, *40*, D420–D427. [CrossRef]
27. Golubkov, V.V.; Dymshits, O.S.; Petrov, V.I.; Shashkin, A.V.; Tsenter, M.Y.; Zhilin, A.A.; Kang, U. Small-Angle X-Ray Scattering and Low-Frequency Raman Scattering Study of Liquid Phase Separation and Crystallization in Titania-Containing Glasses of the ZnO–Al_2O_3–SiO_2 System. *J. Non Cryst. Solids* **2005**, *351*, 711–721. [CrossRef]
28. Alekseeva, I.; Baranov, A.; Dymshits, O.; Ermakov, V.; Golubkov, V.; Tsenter, M.; Zhilin, A. Influence of CoO Addition on Phase Separation and Crystallization of Glasses of the ZnO–Al_2O_3–SiO_2–TiO_2 System. *J. Non Cryst. Solids* **2011**, *357*, 3928–3939. [CrossRef]
29. Ghose, S.; Choudhury, N.; Chaplot, S.L.; Pal Chowdhury, C.; Sharma, S.K. Lattice Dynamics and Raman Spectroscopy of Protoenstatite $Mg_2Si_2O_6$. *Phys. Chem. Miner.* **1994**, *20*, 469–477. [CrossRef]
30. Okuno, M.; Zotov, N.; Schmücker, M.; Schneider, H. Structure of SiO_2–Al_2O_3 Glasses: Combined X-Ray Diffraction, IR and Raman Studies. *J. Non Cryst. Solids* **2005**, *351*, 1032–1038. [CrossRef]

materials

MDPI

Article

In Search of Ultra-Black Ceramic Pigments Using Microwaves: Delafossite Cuprates $CuMO_2$ (M = Mn, Fe, Cr)

Guillermo Monrós [1,*], Vicente Esteve [1], Carolina Delgado [1], Guillem Monrós-Andreu [2] and Mario Llusar [1]

[1] Department of Inorganic and Organic Chemistry, University Jaume I, 12071 Castelló de la Plana, Castelló, Spain; estevev@uji.es (V.E.); caroladelg@gmail.com (C.D.); mllusar@uji.es (M.L.)

[2] Department of Mechanic Engineering and Construction, University Jaume I, 12071 Castelló de la Plana, Castelló, Spain; gmonros@uji.es

* Correspondence: monros@uji.es

Abstract

Cuprate delafossite phases such as $CuMnO_2$ (crednerite) and $CuFeO_2$, as well as iron- and manganese-doped mcconnellite composites, were investigated as candidates for producing intense black ceramic pigments via conventional solid-state synthesis. Both electric kiln and fast dielectric (microwave) firing methods were employed, with mcconnellite ($CuCrO_2$) used as a reference pigment. Microwave firing led to a marked improvement in sample blackness compared to conventional electric firing. Among the delafossite phases, only mcconnellite subjected to microwave-assisted firing (R_{Vis} = 1.40%, corresponding to 98.60% visible light absorption) emerges, pending further optimization, as a promising candidate for an ultra-black ceramic pigment (R_{Vis} < 1%) under optimized glaze conditions (ZnO-free) and a firing temperature of 1000 °C. Considering the pigments in powder form, microwave-fired crednerite (R_{Vis} = 4.85%, 95.15% absorption) and iron- and iron–manganese-doped mcconnellite composites (R_{Vis} = 3.27% and 3.23%, respectively) appear as potential candidates for deep-black pigments (R_{Vis} < 3%), benefiting from the composite effect between the delafossite phase and the associated chromium spinel. Moreover, microwave-fired crednerite also demonstrates noteworthy potential for deep-black coloration in glazed samples (R_{Vis} = 4.27%, 95.73% absorption).

Keywords: ceramic pigment; delafossite; microwaves; blackness; ultra-black; ceramic glaze

1. Introduction

Discovered by Friedel in 1873, the mineral $CuFeO_2$ [1], named *delafossite* in honour of the French mineralogist and crystallographer Gabriel Delafosse, gave its name to the ABO_2 oxide family (delafossites), which have recently attracted growing scientific interest [2–4], particularly since Kawazoe et al. demonstrated the coexistence of optical transparency and p-type conductivity in $CuAlO_2$ [5,6]. Delafossite cuprates, with the general formula ABO_2 (A^+ = Cu; B^{3+} = Mn, Fe, Cr), possess a layered structure composed of BO_6 octahedral sheets linked by linearly coordinated O–A^+–O dumbbells. Pabst (1939) first described the trigonal delafossite structure (space group *R-3m*), while Köhler and Jansen later reported the equivalent hexagonal modification (space group $P6_3/mmc$) [7,8]. At room temperature, however, crednerite ($CuMnO_2$) adopts a unique monoclinic layered structure consisting of edge-shared MnO_6 octahedra and linearly coordinated Cu^+ ions. This arrangement is interpreted as a Jahn–Teller distortion of the delafossite structure induced by Mn^{3+}

ions, making crednerite an intriguing derivative of the delafossite family with promising potential in catalytic applications [9].

The commercial sprayable coating Vantablack (a portmanteau of *VANTA*, "Vertically Aligned Nanotube Arrays," and "black") is an ultra-black surface (L*a*b*= 0.2/0.9/−1; visible reflectance R_{Vis} = 0.035%) produced by chemical vapour deposition (CVD). It absorbs up to 99.965% of incident visible light [10,11]. Ultra-black surfaces conceal surface irregularities and textures, eliminate glare, and render objects nearly invisible, making them highly valuable in space science and optical applications; for instance, in enhancing telescope sensitivity and suppressing stray light in precision optical instruments. In general, an ultra-black surface is defined as one that absorbs more than 99% of incident light (R_{Vis} < 1%) [10,11]. By contrast, commercial black pigments typically exhibit R_{Vis} values below 5%, while Industry Tint Reference Black (ITRB) coatings reach approximately $R_{Vis} \approx 3\%$. Therefore, a deep-black surface may be defined as one that absorbs more than 97% of visible light, corresponding to R_{Vis} < 3%.

In nature, several organisms exhibit ultrablack pigmentation as an adaptive trait shaped by selective pressures within their environments. This pigmentation enhances the visibility of sexually selected colours in peacock spiders and birds-of-paradise, provides thermoregulatory or camouflage advantages to vipers, and aids butterflies in evading predators. Such anti-reflective mechanisms based on ultrablack pigmentation play a crucial role in survival and reproductive success [12].

To produce ultrablack coloration, organisms have evolved sophisticated mechanisms centred on eumelanin (from the Greek *εὺ μέλας*, meaning "true black"). Eumelanin, the black component of the melanin pigment family, is synthesized through melanogenesis; a biochemical process involving the oxidation of the amino acid tyrosine to 5,6-dihydroxyindole, followed by polymerization [13]. However, while eumelanin can generate a deep black colour, it may still exhibit surface gloss, as seen in shiny fur or feathers. In contrast, ultrablack coloration lacks gloss and instead appears matte, owing to hierarchical surface microstructures that modulate light interactions and minimize reflection. Similarly, the microstructure morphology of Vantablack coatings is critical for achieving their exceptional light-absorbing properties [10,11].

Table 1 reports a comparison of black colours published in the recent literature. Commercial black pigments are derived from cobalt ferrochromite (DCMA 13-40-9) and nickel ferrochromite (DCMA 13-50-9), though it is easy to find industrial pigments using other cations, such as Mn. The usual L*a*b* parameters of ferrochromite commercial powder are around 41.0/1.0/1.0 that improve when dispersed in glaze (5 wt%) to values around 33.0/1.0/0.0 with an estimated reflectance in visible range R_{Vis} of 3.5 and 4.5% in powder and glaze, respectively [14,15].

Table 1. Comparison of black ceramic pigments published in the recent literature.

		Powder		Glaze	
	Reference	**L*a*b***	**$R_{Vis}/R_{NIR}/R$ (%)**	**L*a*b* wt%(T°C)**	**$R_{Vis}/R_{NIR}/R$ (%)**
Vantablack	[10]	0.2/0.9/−1	0.035/-/-	--	--
Carbon Black	[11]	20.2/0.1/0.1	3/3/3	--	--
Ferrochromite $(Co,Ni)(Fe,Cr)_2O_4$	[13,14]	41.0/1.0/1.0	3.5/-/-	33.0/1.0/0.05 5(1080)	4.5/-/-
$CuCr_2O_4$	[16]	40.5/0.1/−0.1	4.2/3.4/3.8	26.3/−0.3/−0.5 0.5(800)	3.8/4.5/4.1
$CuCrO_2$	[17]	36.5/−4.8/−0.8	9.9/11.2/10.5	11.3/−0.6/−3.05 5(1000)	1.4/1.9/1.6
Mn,Ni-$CuFe_5O_8$	[18]	21.36/1.1/−0.24		16.95/−0.18/−0.12	

Table 1. *Cont.*

	Powder			Glaze	
	Reference	L*a*b*	$R_{Vis}/R_{NIR}/R$ (%)	L*a*b* wt%(T°C)	$R_{Vis}/R_{NIR}/R$ (%)
$Fe_{1.2}Cr_{0.8}O_3$	[19]	42.3/−0.3/0.6	4/23/14	34.1/2.6/−3.75 5(1000)	4/17/10
$YMnO_3$	[19]	27.7/−1.2/−3.5	3/39/19	31.9/1.6/1.15 5(1000)	7/31/18
$Sr_4CuMn_2O_9$	[19]	43.9/7.2/6.3	7/51/29	57.9/2.8/0.25 5(1000)	25/41/33
$Sr_2(Mg_{0.5}Mn_{0.5})Ge_2O_7$	[19]	31.8/−0.1/−8.3	5/32/17	49.6/12.8/13.75 5(1000)	25/68/44

In a recent study [11], the authors investigated the potential to develop intense black ceramic pigments comparable to, or surpassing, carbon black (L*a*b* = 20.2/0.1/0.1; visible, near-infrared, and total reflectance values $R_{Vis}/R_{NIR}/R$ = 3/3/3%). Carbon black is widely regarded as the reference standard for black inorganic pigments used in paints; however, its poor thermal stability makes it unsuitable for high-temperature ceramic applications.

This goal could be achieved using various cuprates based on spinels, such as $CuCr_2O_4$ [16] or mcconnellite [17]. The tetragonal distorted spinel $CuCr_2O_4$ improves the black colour of ferrochromites, showing the powder L*a*b* = 40.5/0.1/−0.1 and $R_{Vis}/R_{NIR}/R$ = 4.2/3.4/3.8, which improves when was 0.5 wt% glazed in soda-lime glass (800 °C) (L*a*b* = 26.3/−0.3/−0.5 and $R_{Vis}/R_{NIR}/R$ = 3.8/4.5/4.1), reflecting the 96.2% of visible light incident. Likewise, exploring the effect of microwave synthesis on cuprates, using dielectric firing assisted by microwaves (60 min at 800 W), mcconnellite-synthesized pigment powder shows a good black shade (L*a*b* = 36.5/−4.8/−0.8 and $R_{Vis}/R_{NIR}/R$ = 9.9/11.2/10.5), and 5 wt% glazed in a zinc-free ceramic glaze with maturation point at 1000 °C produces an intense coloured black glaze surface with L*a*b* = 11.3/−0.6/−3.0, associated with $R_{Vis}/R_{NIR}/R$ = 1.4/1.9/1.6, thus reflecting 98.6% of the visible light, which borders on ultrablack behaviour. The microstructure of powder prepared by fast dielectric firing and the nature of the glaze enhances the black yield of the glassy coating. Likewise, the presence of zinc in the glaze composition should be avoided, because it reacts with the pigment to produce high stability zinc spinel such as gahnite $ZnAl_2O_4$ or $ZnCr_2O_4$ and exsolve copper (II) in the glassy matrix, creating a shift in colour to green-blue shades [16,17]. Other cuprates such as $CuFe_5O_8$ are codoped with 0.2 molar of MnO_2 and 0.75 molar of Ni_2O_3; using 6.25% of MgO as a mineraliser produces black powders of L*a*b* = 21.36/1.10/−0.24 and 5 wt% when glazed at 1000 °C (with 5.12 wt% of ZnO in its composition). This improves the black colour to L*a*b* = 16.95/−0.18/−0.12, which is lower than that obtained by the simple mcconnellite but better than that of $CuCr_2O_4$ and ferrochromite spinels [18]. Four other important black pigments, such as Cr–hematite to Mn–melilite, were utilized for glazes in [19] and are also included in Table 1. The chromium-doped hematite with R_{Vis} of 4 and 4% in powder and glaze at 1000 °C, respectively, and the perovskite $YMnO_3$ with R_{Vis} of 3 and 7% in powder and glaze at 1000 °C, respectively, stood out. Therefore, from the literature data, delafossite cuprates appear as possible candidates to produce ultrablack ceramic pigments.

In this study, the use of other delafossite cuprates, including $CuMnO_2$ and $CuFeO_2$, as well as iron- and manganese-doped mcconnellite composites, was investigated as potential candidates for producing intense black ceramic pigments. The materials were synthesized via conventional solid-state methods and fired using both electric kilns and fast dielectric (microwave) furnaces, with mcconnellite ($CuCrO_2$) employed as the reference pigment. To the best of our knowledge, the synthesis, characterization, and application of cuprate delafossites (crednerite, delafossite, and mcconnellite) as ceramic pigments in glazes aimed at achieving ultrablack (R_{Vis} < 1%) or deep-black (R_{Vis} < 3%) surfaces have not previously been reported.

2. Materials and Methods

Delafossite $CuMO_2$ (M = Mn,Fe,Cr) pigments were synthesized by solid-state method using tenorite CuO, eskolaite Cr_2O_3, bixbyite Mn_2O_3, and hematite Fe_2O_3 oxides as precursors, with a particle size of 0.3–5 μm (supplied by QUIMIALMEL SA, Castelló, Spain, 99.8 wt% purity). The precursors were mechanically homogenized in an electric grinder (20,000 rpm) for 5 min and then fired at 1100 °C for 3 h in electric kiln and also using fast microwave-assisted dielectric firing [20,21].

This hybrid microwave-assisted heat treatment of solid reactant mixtures is conducted in a closed kiln internally coated with a susceptor, which functions as an external heater during the initial stage of microwave irradiation [22]. Upon exposure to microwaves, the susceptor rapidly increases in temperature and transfers heat to the reactant mixture by radiation and/or conduction. Once the mixture reaches the critical temperature of the irradiated material, it begins to absorb microwave energy directly and continues heating autonomously.

In this study, dielectric firing was performed using a conventional microwave equipment operating at 900 W and 2.45 GHz, equipped with a cylindrical kiln for preheating the reactant mixture. The kiln (inner dimensions: height 4.5 cm, diameter 13.5 cm) was constructed from aluminosilicate fibreboard (density: 0.3 g cm^{-3}), a material transparent to microwaves with penetration depths of 0.007 m at room temperature and 0.0025 m at 1300 °C. The kiln interior was coated with α-SiC (hexagonal, density: 3.2 g cm^{-3}) serving as a susceptor (Glass Fusing & Cutting Tools, Tianjin, China; product code 1282151), as previously described in [17]. The temperature during firing was estimated using Process Temperature Control Rings, M.E. SCHUPP, Aachen, Germany (PTCR, ETH type, 850–1100 °C). Based on PTCR shrinkage measurements taken with a micrometre, the estimated temperatures reached approximately 950 °C after 20 min, 1050 °C after 30 min, and 1150 °C after 60 min of irradiation.

X-ray Diffraction (XRD) was performed on a Siemens D5000 diffractometer using Cu K_α radiation (10–70°2θ range, scan rate 0.03°2θ, 5 s per step and 40 kV and 20 mA conditions). Additional XRD processes with the addition of 30 wt% of α-Al_2O_3 as an internal measurement pattern (10–70°2θ range, scan rate 0.02°2θ, 10 s per step and 40 kV and 20 mA conditions) were performed for cell parameters estimations. The lattice constants were determined from corrected peak positions using α-Al_2O_3 as standard and least square refinement programmes.

The L*a*b* colour parameters of samples were measured according to the CIE-L*a*b* (Commission Internationale de l'Éclairage) [23], using an X-Rite SP60 spectrometer with standard lighting D65 and a 10° observer. L* measures the lightness (100 = white, 0 = black) and *a** and *b** the chromatic components (−*a** = green, +*a** = red, −*b** = blue, +*b** = yellow).

The tolerance Δ*E** (based on the L*a*b* parameters), is evaluated using Equation (1):

$$\Delta E^* = \sqrt{\Delta L^{*2} + \Delta a^{*2} + \Delta b^{*2}} \quad (1)$$

UV–Vis–NIR spectra of the samples were recorded using a Jasco V-670 diffuse reflectance spectrophotometer. The optical band gaps were estimated using the Tauc method [24]. For the optimized pigments, optical reflectance spectra were collected over the 350–2500 nm range. The total solar reflectance (R), the solar reflectance in the near-infrared range (R_{Nir}), and the solar reflectance in the visible range (R_{Vis}) were evaluated from the UV–Vis–NIR diffuse reflectance spectra. These parameters were calculated as the ratio between the integral of the measured spectral irradiance multiplied by the solar irradiance and the integral of the solar irradiance over the same wavelength interval, namely 350–2500 nm for R, 700–2500 nm for R_{NIR}, and 350–700 nm for R_{Vis}, as expressed in Equation (2):

$$R = \frac{\int_{350}^{2500} r(\lambda)i(\lambda)d\lambda}{\int_{350}^{2500} i(\lambda)d\lambda} \tag{2}$$

where $r(\lambda)$ is the dimensionless spectral reflectance measured from UV–Vis–NIR spectroscopy and $i\lambda$) is the standard solar irradiance ($Wm^{-2}nm^{-1}$), according to the American Society for Testing and Materials (ASTM)'s Standard G173-03 [25] The solar absorption spectrum of the samples, representing their wavelength-dependent absorption relative to solar irradiance, was calculated by evaluating the numerator and denominator of the equation at each wavelength.

The pigmenting capacity of the samples was evaluated by incorporating the pigment into a double-firing frit, zinc-free, with a maturation point of 1000 °C (composition in wt%: SiO_2 72, Na_2O 14, K_2O 1.5, CaO 9, Al_2O_3 2.3; average grain size: 3–20 μm; supplied by Torrecid S.A., l'Alcora, Spain). Glazed samples were prepared on white stoneware tile substrates. The glaze was manually mixed with the pigment and water in a weight ratio of 100:5:40 using an agate mortar and applied to a thickness of approximately 1500 μm using the Doctor Blade technique. A 40% shrinkage of the glaze layer was observed after firing.

Molten glazes interact with pigment particles through dissolution or chemical reactions, potentially leading to pigment degradation or colour alteration. The aggressiveness of the glaze depends on both its chemical composition and the firing temperature. The Dry Color Manufacturers Association classifies pigment powders into three categories [26,27]:

- *Category A* deals with pigments suspended in glass matrixes which require the highest degree of heat stability and chemical resistance to withstand the attack of molten glass.
- *Category B* deals with pigments suspended in plastics and other polymers, which require only moderate heat stability.
- *Category C* deals with pigments suspended in liquid vehicles, which require little, or no heat stability.

Ceramic pigments belong to Category A, showing stability and chemical resistance to molten ceramic glazes: Vantablack and Carbon Black belong to Category C and B, respectively.

In the field of polymers, Carbon Black is the reference standard for deep-black shades. The Industry Tint Reference Black (ITRB) is a standardized Carbon Black material used to evaluate the tinting strength of other Carbon Black pigments in polymers, following ASTM D3265 [28] or ISO 5435 [29] test methods. These methods involve mixing the black standard with zinc oxide and a liquid organic plasticizer. The resulting paste is mulled, homogenized, and spread as a layer of uniform thickness. Its light reflectance is then measured with a reflectometer and compared to a paste prepared with the sample under investigation.

Three specific metrics are used to quantify the degree of blackness based on tristimulus values ((X, Y, Z) for the sample and (Xn, Yn, Zn) for the reference): (a) blackness ($My = 100 \log_{10}(Yn/Y)$) measures the overall degree of blackness and is directly related to reflectance; (b) jetness or darkness ($Mc = 100 [\log_{10}(Xn/X) + \log_{10}(Yn/Y) - \log_{10}(Zn/Z)]$) reflects the colour-dependent black value, influenced by primary particle size, surface area, and dispersion in the polymer; and (c) undertone ($dM = Mc - My$) indicates whether the black exhibits a brown-reddish undertone ($dM < 0$), is neutral ($dM = 0$), or shows a bluish undertone ($dM > 0$), the latter often being preferred [30,31]. These metrics, however, are not directly applicable to ceramic pigments, which must remain stable at high temperatures when incorporated into ceramic glazes.

3. Results and Discussion

3.1. Characterization of Powders: Effect of Microwaves Firing

Figure 1 and Table 2 summarize the properties of the powders (fired in an electric kiln and via microwave irradiation for 30 min at 900 W) and of samples containing 5 wt% pigment glazed in a ZnO-free frit with a maturation temperature of 1000 °C.

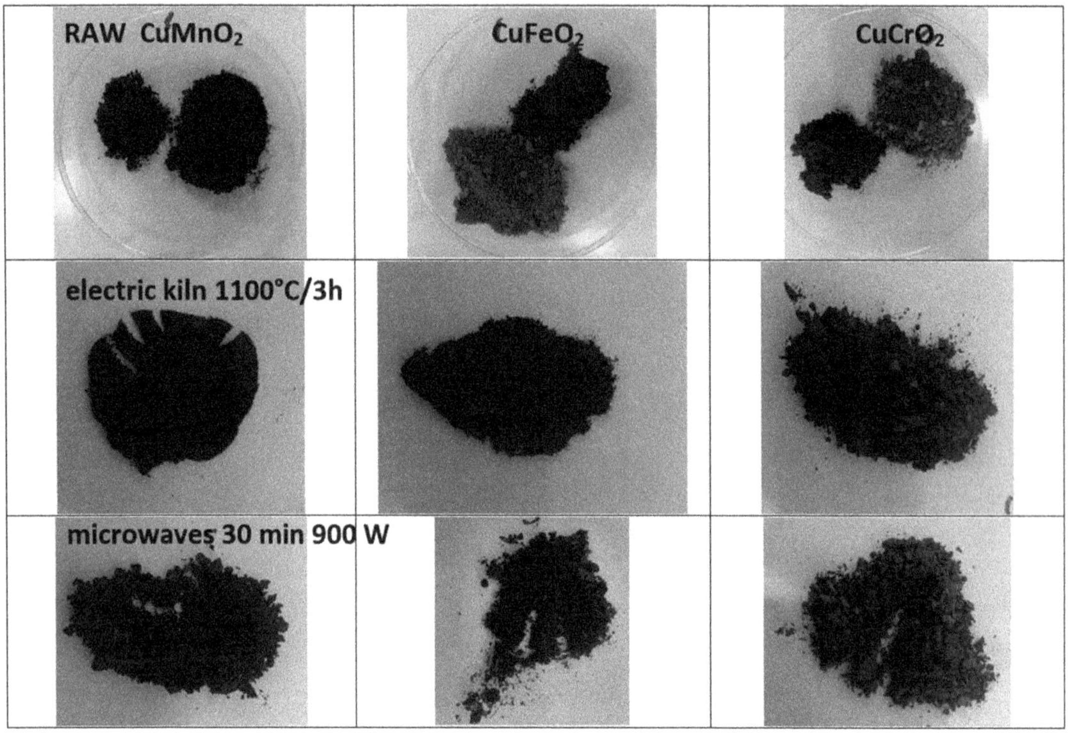

Figure 1. Powders before mixture and fired in electric kiln and with microwaves at 900 W for 30 min).

All powders fired in the electric furnace exhibited a dark black appearance. However, subtle undertones were observed: manganese crednerite displayed a reddish-brown hue ($a^* > 0$), iron delafossite powders exhibited a bluish undertone ($b^* < 0$), and mcconnellite showed a greenish undertone ($a^* < 0$). The visible-range reflectance (R_{Vis}) follows the sequence 1MnCE < 3CrCE < 2FeCE, corresponding to a blackness order of 1MnCE > 3CrCE > 2FeCE. This trend is consistent with the increasing band gap (in the NIR range) and decreasing ΔE^* values with increasing blackness (Table 2).

For microwave-fired powders, the R_{Vis} sequence differs (1MnMW < 3FeMW < 2CrMW) and is associated with increasing ΔE^*, although the band gap values are comparable to those of the electrically fired samples and follow the same sequence. Microwave firing reduced R_{Vis} for the Mn and Fe samples, enhancing their blackness, whereas the Cr sample exhibited a slightly higher R_{Vis} under microwave firing compared to the electric-fired sample.

The XRD diffractograms of samples are shown in Figure 2 and the crystalline phases identified are summarized in Table 3.

Table 2. Characterization of black samples: fired at 1100 °C for 3 h in an electric kiln (CE), fired at 900 W for 30 min in a microwaves kiln (MW). ΔE* is referred to as Carbon Black (L*a*b* = 20.2/0.1/0.1).

SAMPLE	L*a*b*	$R_{Vis}/R_{NIR}/R$ (%)	E_g (eV)	ΔE*
POWDER				
1MnCE	27.32/1.18/0.40	5.47/6.06/5.73	1.31	7.2
2FeCE	31.14/−0.29/−2.47	6.87/7.10/6.98	1.42	11.2
3CrCE	28.66/−4.17/1.17	6.10/8.26/7.03	1.40	9.5
1MnMW	25.23/3.25/5.18	4.85/6.36/5.51	1.31	7.8
2FeMW	29.18/0.56/−1.70	6.45/13.40/9.52	1.47	9.2
3CrMW	31.05/−2.39/−0.15	6.98/7.73/7.31	1.40	11.1
GLAZED ELECTRIC KILN 5 wt% with glaze 1000				
1MnCE	32.49/−1.93/−6.73	7.23/13.11/9.88	1.21	14.2
2FeCE	25.53/0.60/−5.93	5.02/13.99/9.01	1.21	8.1
3CrCE	24.70/−0.77/0.32	4.40/4.81/4.58	1.45	4.6
1MnMW	26.80/−0.63/−4.90	5.10/12.51/8.41	1.31	8.3
2FeMW	31.25/−0.12/−6.04	7.14/15.40/10.82	1.16	12.6
3CrMW	23.52/−0.02/−0.98	1.41/2.14/1.77	1.38	3.5
GLAZED MICROWAVES 5 wt% with glaze 1000				
1MnCE	31.10/−0.77/1.77	6.37/8.50/7.32	1.50	11.1
2FeCE	28.60/−2.63/1.25	5.70/8.79/7.06	1.41	8.9
3CrCE	31.91/−4.67/3.88	5.95/9.42/7.50	1.80	13.2
1MnMW	24.64/0.02/0.40	4.27/5.93/5.01	1.51	4.5
2FeMW	34.30/−5.94/2.31	6.99/10.66/8.64	1.48	15.4
3CrMW	24.25/−1.41/−0.23	2.63/6.63/5.42	1.49	4.3

In the case of crednerite ($CuMnO_2$), the spinel phase $CuMn_2O_4$ crystallizes alongside monoclinic crednerite under conventional electric kiln firing. However, when synthesized using a microwave kiln (1MnMW), crednerite exhibits very weak XRD peak intensities. Residual peaks corresponding to CuO and Mn_2O_3 are also detected in both samples. The observed phase composition aligns with the Cu–Mn–O phase equilibria [32], where crednerite is metastable at room temperature and its formation is strongly dependent on the heat treatment of the oxide mixture. As reported in [33–35], the XRD pattern of $CuMnO_2$, prepared via low-temperature exchange reactions and essentially free of impurities, can be indexed to a monoclinic crednerite structure (space group *C2/m*) Thermogravimetric analysis (TGA) reveals an oxygen uptake, resulting in a reversible transition to the oxygen-rich spinel $Cu_xMn_{2-x}O_4$ ($x \approx 1.03$). Subsequent firing above 940 °C induces oxygen loss, converting the oxide back to $Cu_{1.1}Mn_{0.9}O_2$, with residual spinel remaining as an impurity. At 1150 °C, $CuMnO_2$ undergoes peritectic melting, and XRD indicates the coexistence of cubic spinel and CuO/Cu_2O phases. Consistently, sample 1MnCE was observed to melt after firing at 1200 °C for three hours.

The indexing of monoclinic crednerite $CuMnO_2$ (*C2/m* ICDD 50-0860) diffraction peaks for 1MnCE indicates a unit-cell volume slightly larger than previously reported values (V = 93.5(3) Å^3, compared with 92.5 Å^3 by Trari et al. [33] and 92.1 Å^3 in the literature [34,35]) (Table 4). $CuMnO_2$ forms a solid solution of the general composition $Cu^{+}{}_{1-x-2y}Cu^{2+}{}_{y}Mn^{3+}{}_{1-x}Mn^{4+}{}_{x}O_2$, with cationic defects located within O–Cu–O dumbbells. In copper manganite, CO conversion is governed by the redox equilibrium $Cu^{+} + Mn^{3+} \longleftrightarrow Cu^{2+} + Mn^{4+}$, where CO chemisorbs on Cu^{2+}/Mn^{4+} sites and O_2 on

Cu^+/Mn^{3+} sites [36]. The ionic radius of Cu^+ (0.60 Å, CN = 2) in O–Cu–O dumbbells is larger than that of Cu^{2+}, whereas Mn^{3+} in octahedral coordination (0.72 Å for low spin; 0.785 Å for high spin) is larger than Mn^{4+} (0.67 Å) [37]. Consequently, the substitution of Cu^{2+} into O–Cu–O sites, accompanied by the reduction of Mn^{4+} to Mn^{3+} in octahedra, leads to lattice expansion, producing a net dilation consistent with the experimentally observed unit-cell volume. The indexing of peaks of spinel $CuMn_2O_4$ as cubic phase (*Fd3m* ICDD 74–2422) for 1MnCE and 1MnMW gives a cell parameter of a = 8.308(1) Å, which is very close to the literature data [38] (Table 4).

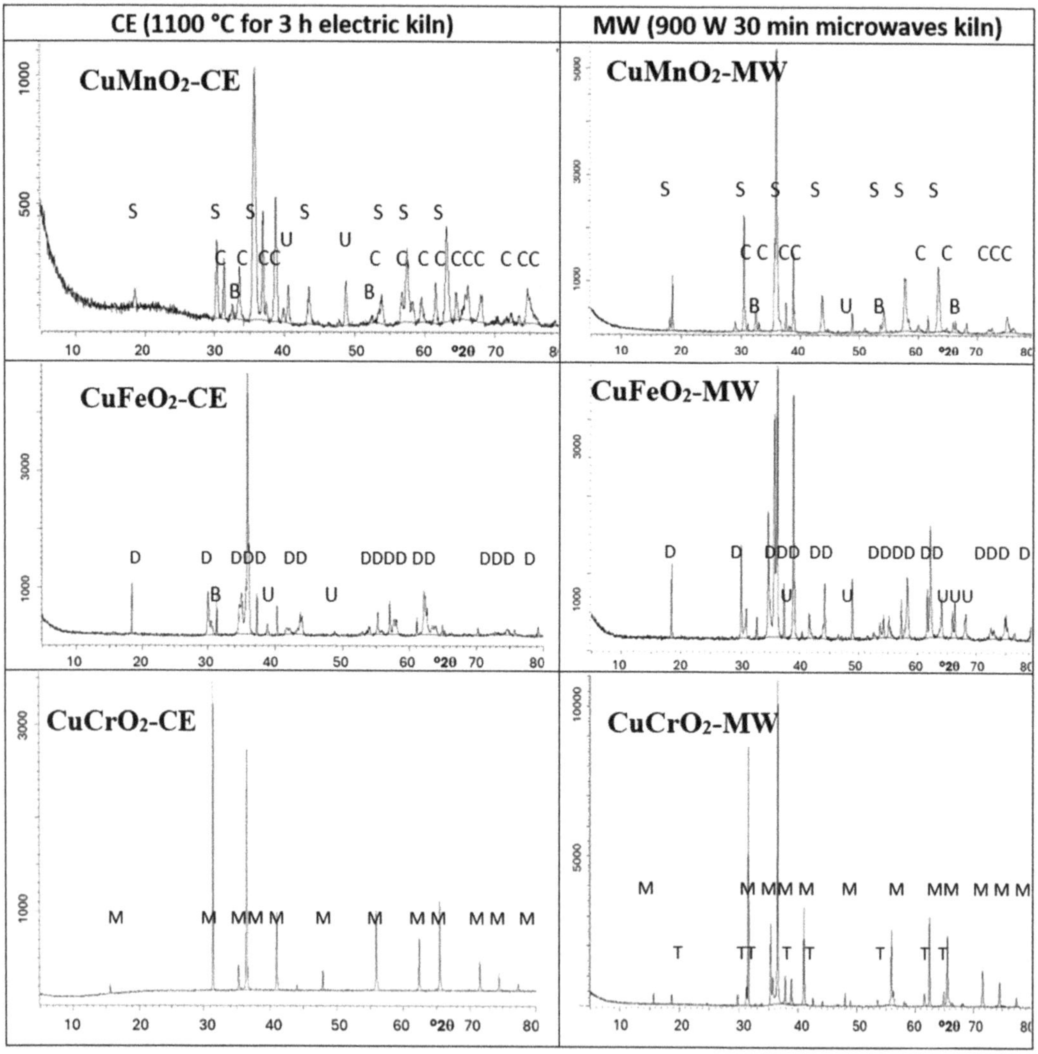

Figure 2. XRD of powders. Crystalline phases: S ($CuMn_2O_4$), C ($CuMnO_2$ crednerite), D ($CuFeO_2$ delafossite), M ($CuCrO_2$ mcconnellite), U (CuO), B (Mn_2O_3), and T ($CuCr_2O_4$). (the abscissa of each diffractogram is °2θ (2 degrees theta) and the ordinate is "counts").

Table 3. XRD results. Crystalline phases: S ($CuMn_2O_4$), C ($CuMnO_2$ Crednerite), D ($CuFeO_2$ Delafossite), M ($CuCrO_2$ mcconnellite), U (CuO), B (Mn_2O_3), and E (eskolaite Cr_2O_3), T ($CuCr_2O_4$). Intensity: vs (very strong), s (strong), m (medium), W (weak), and vw (very weak).

SAMPLE	Crystalline Phases
11MnCE	S (vs) C (m) U, B (vw)
2FeCE	D (vs) U, B (vw)
3CrCE	M (s) E, U (vw)
1MnMW	S (s) C (w) U, B (vw)
2FeMW	D (vs) U (s)
3CrMW	M (s) T (vw)

Table 4. Cell parameters of identified crystalline phases.

Sample	Identified Crystalline Phases		References
$CuMnO_2$-CE	$CuMn_2O_4$ cubic (ICDD 74–2422) a = 8.308(1) Å	$CuMnO_2$ monoclinic (ICDD 50-0860) β = 104.08° a = 5.677(1) Å b = 2.801(2) Å c = 6.063(2) Å V = 93.5(3) Å^3	[34] $CuMnO_2$ monoclinic (*C2/m* ICDD 50-0860) β = 104.03(3)° a = 5.592(3) Å b = 2.883(1) Å c= 5.892(3) Å V = 92.1(1) Å^3
$CuMnO_2$-MW	$CuMn_2O_4$ cubic ICDD 74–2422) a= 8.308(1) Å	--	[35] β = 104.02(3)° a = 5.596(2) Å b = 2.880(1) Å c = 5.899(2) Å [38] $CuMn_2O_4$ cubic *Fd3m* ICDD 74-2422 a = 8.308 Å
$CuFeO_2$-CE	$CuFeO_2$ (ICDD-75-2146) a = b = 3.093(1) Å c = 17.109(1) Å V = 136.0(3) Å^3	--	[39] (ICDD-75-2146) a = b = 3.04 Å c = 17.15 Å V = 136.8668 Å^3
$CuFeO_2$-MW	$CuFeO_2$ (ICDD-75-2146) a = b = 3.002(1) Å c = 17.392(1) Å V = 135.8(3) Å^3	--	
$CuCrO_2$-CE	$CuCrO_2$ (ICDD 39-0247) a = b = 2.968(5) Å c = 17.002(3) Å V= 129.7(4) Å^3	--	[40,41] ICDD 39-0247 *a* = 2.9480(2) Å *c* = 17.033(2) Å V = 127.9731 Å^3
$CuCrO_2$-MW	$CuCrO_2$ (ICDD 39-0247) a = b = 2.967(4) Å c = 17.021(2) Å V = 129.9(5) Å^3	--	

The different compositions of the samples, with a composite of spinel and crednerite undergoing the electric firing and the spinel as the only phase undergoing microwave firing, explains the slight decrease in R_{Vis} in the microwave-fired sample (5.47 and 4.85% for 1MnCE and 1MnMW, respectively), indicating the lower black colour of the crednerite $CuMnO_2$ than that of the $CuCr_2O_4$ spinel. However, additional factors may also be involved, such as the so-called "composite effect" arising from pigments composed of different elements and compounds. This effect can modify the reflection, absorption, and scattering of light, thereby generating distinct colour characteristics. Furthermore, the size and morphology of pigment particles influence their optical behaviour: larger particles generally enhance light scattering, whereas smaller particles promote greater light absorption. In this case, the composite effect may contribute to the enhanced blackness of 1MnCE, while the smaller particle size of 1MnMW) likely intensifies its black appearance.

The combined influence of these factors accounts for the slightly increased blackness observed in 1MnMW.

For delafossite $CuFeO_2$, both samples display intense peaks corresponding to the delafossite phase (space group *R-3m*). In contrast to the electrically fired sample (2FeCE), which exhibits only very weak peaks of unreacted oxides, the microwave-fired sample (2FeMW) shows pronounced peaks corresponding to tenorite (CuO). The indexing of the delafossite peaks (ICDD 75-2146; Table 4) reveals lattice parameters slightly lower than those reported in the literature [39]. Specifically, the cell volume of the delafossite in 2FeCE is marginally smaller than that of 2FeMW (V = 136.0(3) and 135.8(3) $Å^3$ for 2FeMW and 2FeCE, respectively, compared to 136.8668 $Å^3$ in reference [39]).

The composite effect in sample 2FeMW (CuO intimately mixed with $CuFeO_2$), along with the possibly lower blackness of delafossite compared with tenorite CuO, can explain the decrease in R_{Vis} for the microwave-fired powder (6.87% and 6.45% for 2FeCE and 2FeMW, respectively).

For the reference mcconnellite, XRD analysis shows that $CuCrO_2$ crystallizes as the predominant phase in both electrically fired (3CrCE) and microwave-fired (3CrMW) samples. In the electrically fired 3CrCE, only mcconnellite peaks are observed. In contrast, the microwave-fired 3CrMW exhibits weak peaks correspond to a tetragonally distorted spinel, $CuCr_2O_4$. The indexing of the mcconnellite peaks (space group *R3m*, ICDD 39-0247; Table 4) indicates lattice parameters slightly higher than those reported in the literature [40,41]. Specifically, the cell volume of mcconnellite in 3CrMW is marginally larger than in 3CrCE (V = 127.9731 $Å^3$ in reference [40,41], 129.7(4) $Å^3$ for 3CrMW, and 129.9(5) $Å^3$ for 3CrCE).

The high content of black-coloured mcconnellite in the 3CrCE sample explains the slight increase in R_{Vis} (i.e., a decrease in blackness) in the microwave-fired sample (R_{Vis} = 6.10% and 6.98% for 3CrCE and 3CrMW, respectively)

Figure 3 presents the UV–Vis–NIR diffuse reflectance spectra of the powders. Both manganese-containing samples exhibit nearly featureless spectra, with only a very weak band at 310 nm and an additional reflectance minimum at 430 nm for the microwave-fired 1MnMW sample. Similarly, the iron delafossite powder (2FeCE), which consists predominantly of $CuFeO_2$, shows a mostly featureless spectrum with a weak minimum at 310 nm. In contrast, the microwave-fired 2FeMW powder displays a pronounced increase in reflectance in the NIR region, consistent with the presence of CuO detected by XRD. The reference mcconnellite exhibits largely featureless spectra in both firing conditions, with weak minima (corresponding to absorbance maxima), observed at 430, 620, 680, and 1200 nm. These features are attributed to Cr^{3+} (d^3) electronic transitions in octahedral coordination: $^4A_2g(^4F) \rightarrow {}^4T_1g(^4F)$ at 430 nm, $^4A_2g(^4F) \rightarrow {}^4T_2g(^4F)$ at 680 nm, and $^4A_2g(^4F) \rightarrow {}^2E\_g(^2G)$ at 1200 nm. The absorption at 620 nm is assigned to low-coordinated Cu^+ ions in the O–Cu^+–O dumbbells of the mcconnellite structure [42].

Figure 4 presents the solar reflectance spectra and Tauc plots of the powders. The band gaps are identical for the crednerite and mcconnellite samples under both electric- and microwave-firing conditions (1.31 eV and 1.40 eV, respectively). In contrast, the iron delafossite samples exhibit different band gaps due to the presence of CuO in the microwave-fired 2FeMW sample. The solar spectra highlight the strong absorption of solar radiation by all intense black powders across both the visible and near-infrared regions.

Figure 5 shows optical images of the powders at two magnifications (×100 and ×500) alongside SEM micrographs. The ×100 optical images reveal that manganese crednerite exhibits a brown undertone, iron delafossite a bluish undertone, and mcconnellite a greenish undertone, consistent with the L*a*b* colour parameters. The analysis of the ×500 optical images and SEM micrographs allows the estimation of particle size, indicating that microwave firing produces smaller particles in all cases compared to electric firing. Particle size has a moderate influence on the blackness of the powders: R_{Vis} decreases slightly for manganese and iron powders and for mcconnellite in the microwave-fired samples. The effect of particle size is expected to be more pronounced in glazed samples, as smaller particles increase the number of colour points per unit volume within the glassy matrix, enhancing the perceived blackness.

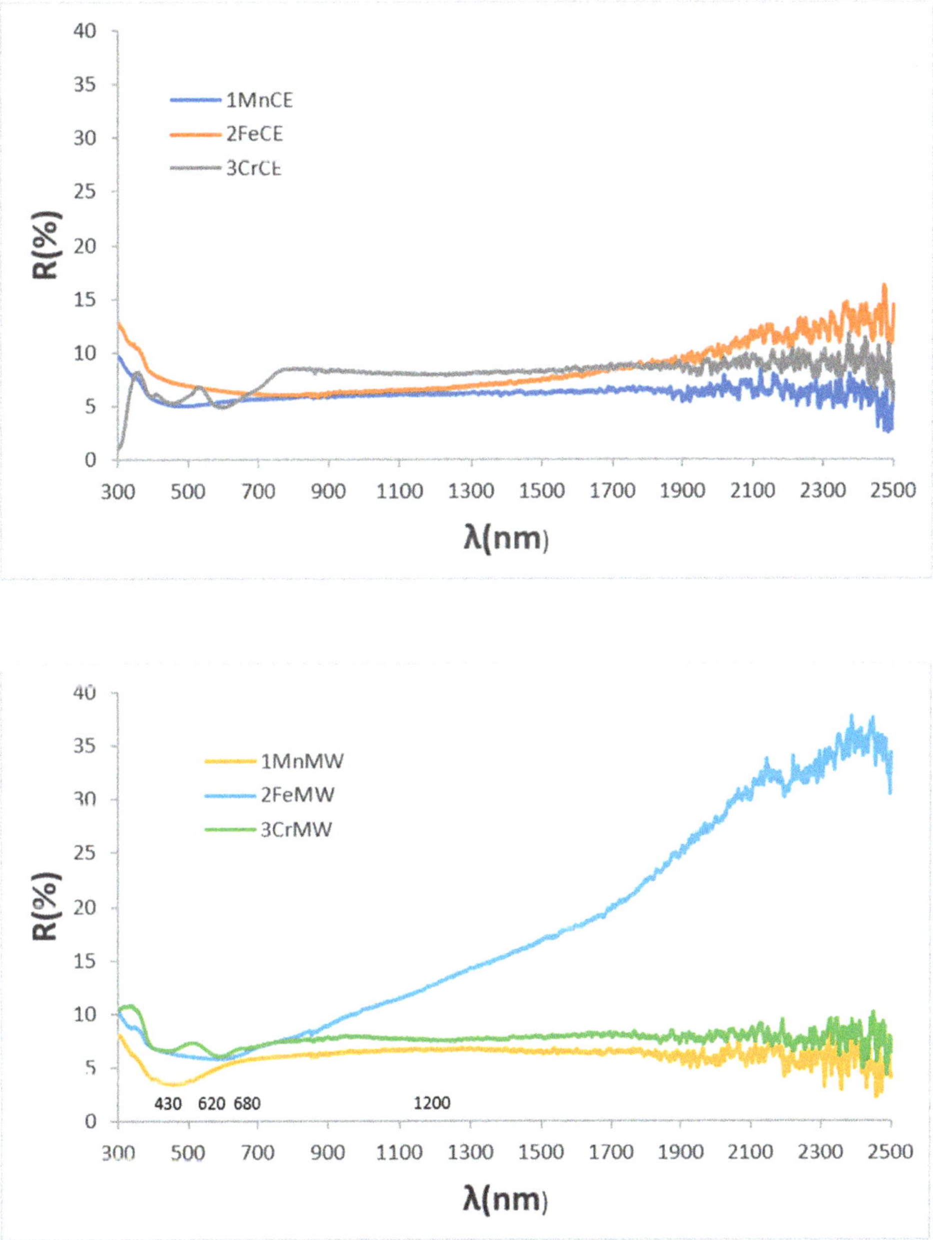

Figure 3. UV–Vis–NIR reflectance spectra of powders.

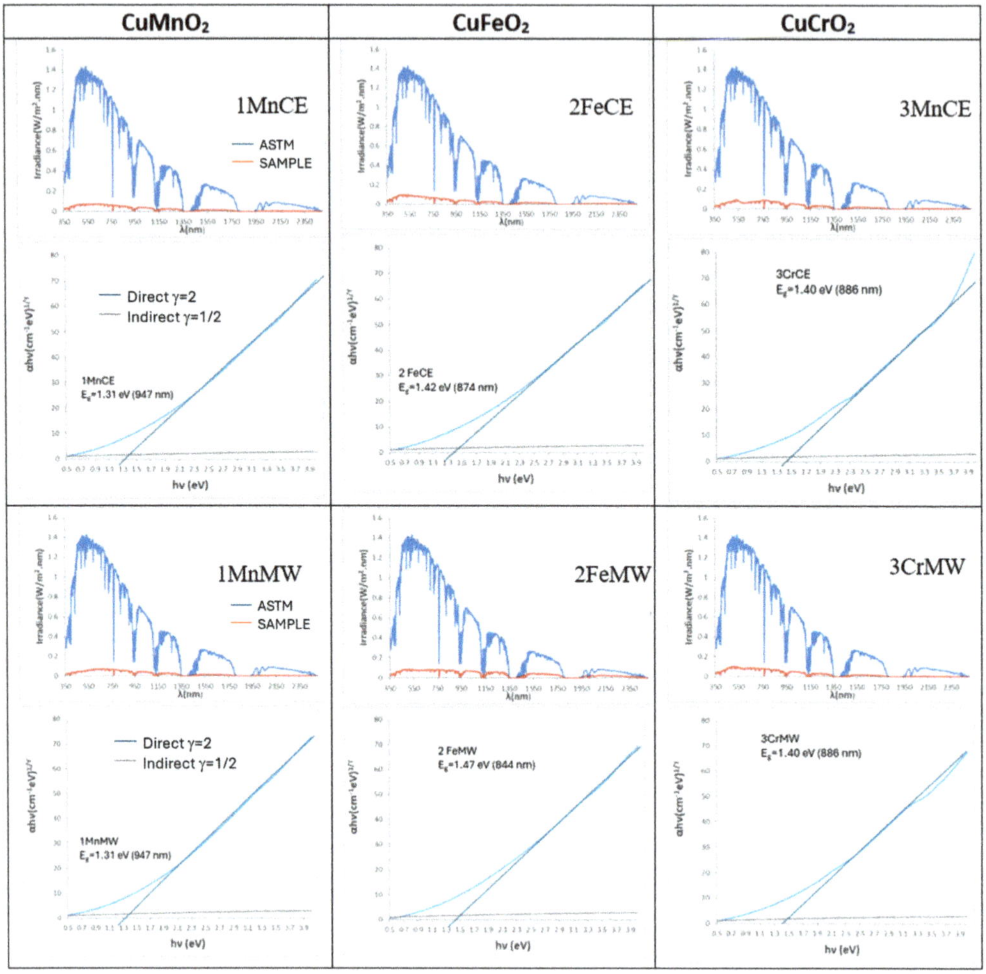

Figure 4. Tauc's plot and solar spectra of powders.

3.2. Characterization of 5 wt% Glazed Samples in Double Firing Frit with Maturation Point at 1000 °C, Zinc-Free, Using Electric Firing for Glazing

Figure 6 and Table 2 show the characteristics of 5 wt% glazed powders in double firing frit with maturation point at 1000 °C, zinc-free, using an electric kiln for glazing.

All electrically fired glazes appear dark black. Manganese- and iron-containing glazes exhibit a bluish undertone ($b^* < 0$), whereas mcconnellite shows a neutral to greenish hue ($a^* \leq 0$). Among glazes derived from electrically fired powders, R_{Vis} decreases in the sequence 1MnCE < 2CrCE < 3FeCE, reflecting the blackness trend 3CrCE > 2FeCE > 1MnCE and the corresponding decline in ΔE^* (Table 1). Mcconnellite stands out with an intense black tone (R_{Vis} = 4.40, ΔE^* = 4.6) and a higher measured bandgap than the original powder.

For glazes derived from microwave-fired powders, the R_{Vis} sequence follows 2FeMW > 1MnMW > 2CrMW, consistent with the decreasing trends of both ΔE^* and bandgap. Mcconnellite is notable for its deep-black appearance (R_{Vis} = 1.41, ΔE^* = 3.5), accompanied by a reduced bandgap relative to the corresponding powder. Overall, Mn- and Cr-containing glazes from microwave firing exhibit lower R_{Vis} values than their electrically fired counterparts, indicating a deeper black shade. In contrast, the Fe sample shows a higher R_{Vis} under microwave firing, which is attributed to the presence of CuO in the powder.

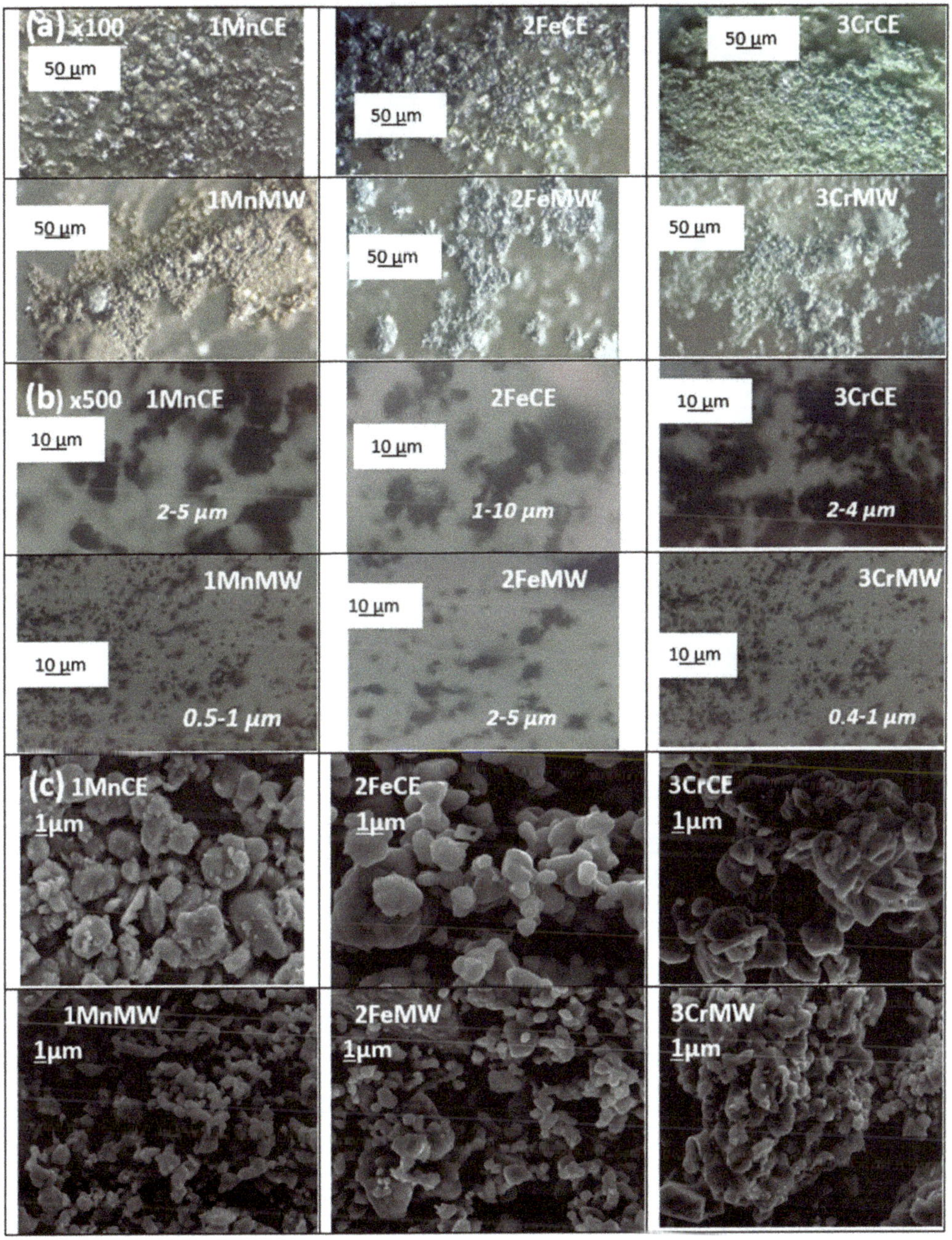

Figure 5. (**a**) Optical images of powders ×100; (**b**) Optical images of powders ×500; (**c**) SEM micrographs of powders (the segment below the value in microns is the corresponding length).

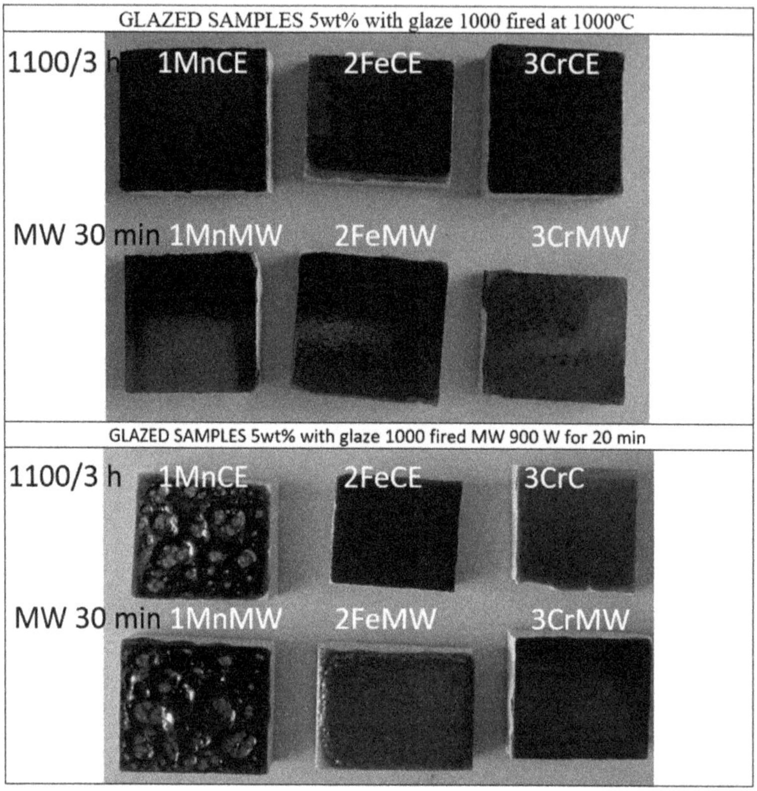

Figure 6. Images of 5 wt% glazed samples in ZnO-free frit of 1000 °C.

Figure 7 presents the XRD diffractograms of the samples, recorded under both backscattering and grazing-incidence conditions (2° incidence angle). X-ray diffraction (XRD) analyses were carried out using a Siemens D5000 diffractometer with Cu Kα radiation over the 2θ range of 10–70°, employing a step size of 0.02° 2θ and a counting time of 10 s per step, under operating conditions of 40 kV and 20 mA. These parameters enable the assessment of pigment particle behaviour dispersed within the amorphous silica glassy matrix, which contributes to the black coloration of the samples.

The X-ray diffraction intensity of crystalline phases detected in glazed samples depends on several factors, including the atomic number (elements with higher atomic numbers possess greater scattering power; in this case, Cr [Z = 24], Mn [Z = 25], and Fe [Z = 26] exhibit comparable scattering ability), the degree of crystallinity (higher crystallinity increases the number of atoms contributing to diffraction), crystal size (larger crystals yield stronger diffraction peaks), and the crystallographic orientation of the crystal relative to the incident X-ray beam [43]. Therefore, the detection of crystalline phases within glazes sharing a similar glass matrix, comparable atomic numbers, and equivalent beam orientation requires that particles exceed certain thresholds of concentration and size. Finely dispersed and/or low-concentration crystalline phases, particularly those affected by glass corrosion, may fall below the detection limit of the XRD technique.

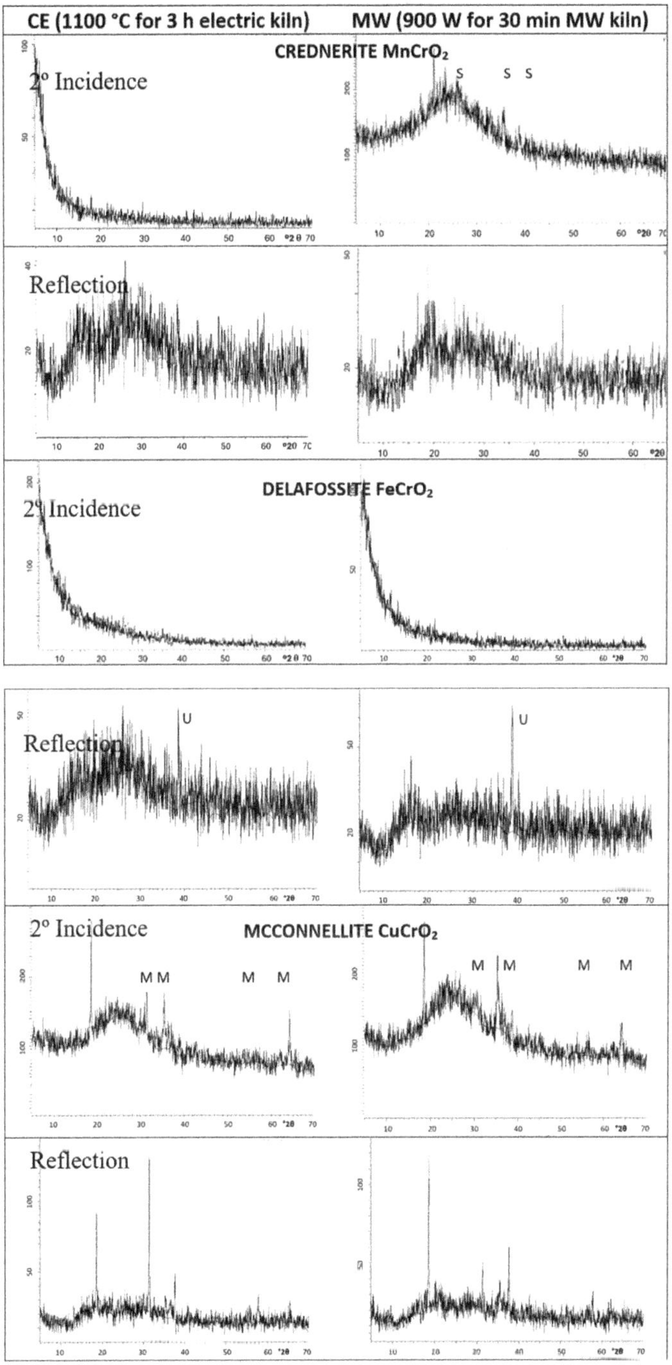

Figure 7. XRD low-incidence angle (2°) and conventional backscattering of 5 wt% glazed samples in frit fired with electric kiln. Crystalline phases: S ($CuMn_2O_4$), M ($CuCrO_2$ McConnellite), U (CuO tenorite). (the abscissa of each diffractogram is °2θ (2 degrees theta) and the ordinate is "counts").

For crednerite ($CuMnO_2$), the grazing-incidence XRD of the glazed microwave-fired powder 1MnMW reveals peaks corresponding to the spinel phase $CuMn_2O_4$. In contrast, no peaks or crystallization halos are detected in the glazed 1MnCE powder. Backscattering XRD shows a double crystallization halo between 20° and 25° 2θ, attributed to the silica glass matrix. Crednerite is not observed after interaction with the molten glaze; thus, the highly stable spinel $CuMn_2O_4$ is identified as the crystalline phase responsible for the black coloration.

For delafossite ($CuFeO_2$), grazing-incidence XRD shows no detectable response. Conventional backscattering XRD displays crystallization halos at ~20° and 25° 2θ, along with a diffraction peak at 39° 2θ, corresponding to the intense [114] reflection plane of tenorite (CuO). This indicates that delafossite decomposes during glazing, and the black coloration cannot be attributed to $CuFeO_2$, but rather to the concentration of CuO. It is well established that Cu^{2+} ions dissolved in glazes derived from copper-based pigments can produce either a green or a blue coloration, depending on their coordination environment: octahedrally coordinated Cu^{2+} in the glassy matrix is associated with green hues, while square-planar coordinated Cu^{2+} surrounded by oxygen atoms in the glass network yields blue tones [44,45]. In the present case, however, the behaviour differs. The 1FeCE powder crystallizes predominantly as delafossite, with weak peaks of CuO phase, whereas the 1FeMW powder exhibits pronounced CuO diffraction peaks. According to the literature, delafossite is thermally unstable and decomposes into CuO at approximately 1070 °C [46]. These results therefore indicate that microwave irradiation accelerates the decomposition of delafossite into CuO. In the glazed samples, however, the low-incidence XRD patterns are nearly identical for both cases, showing the intense [114] reflection plane characteristic of tenorite (CuO) in each. Both glazed samples also exhibit a greenish hue, consistent with the partial dissolution of copper into the glaze. These results indicate that delafossite is unstable and decomposes into CuO within the glassy matrix, following a mechanism analogous to its thermal decomposition at elevated temperatures.

For the reference mcconnellite, XRD confirms the crystallization of $CuCrO_2$ in both glazes prepared from electric- and microwave-fired powders, analyzed by conventional and grazing-incidence XRD. The resistance of mcconnellite to molten glass highlights the high stability and effectiveness of $CuCrO_2$ as a black ceramic pigment.

Figure 8 shows the UV–Vis–NIR reflectance spectra of samples glazed in an electric kiln. Consistent with the XRD results, only mcconnellite, which remains stable against attack by the molten glaze, exhibits a flat reflectance across the Vis–NIR range. In contrast, manganese- and iron-containing delafossites, which are unstable in the glaze medium and crystallize as spinel $CuMn_2O_4$ and CuO, respectively, show increased reflectance in the NIR region, a feature characteristic of potential solar-selective absorber surfaces [17].

Figure 9 shows the solar spectra and Tauc plots of glazed samples prepared by electric firing. Bandgap values vary between samples from electrically and microwave-fired powders, reflecting the influence of different crystallization systems within the glass matrix. The spectra reveal strong solar absorption across the visible and near-infrared ranges, with sample 3CrMW exhibiting particularly high absorption, consistent with its deep-black colour (R_{Vis} = 1.41%), highlighting the effect of crystallization on optical performance.

3.3. Characterization of 5 wt% Glazed Samples in Double Firing Frit with Maturation Point at 1000 °C, Zinc-Free, Using Microwaves Firing for Glazing

Figure 6 and Table 2 summarize the characteristics of 5 wt% glazed powders in a double firing frit with a maturation point of 1000 °C, prepared using a microwave kiln.

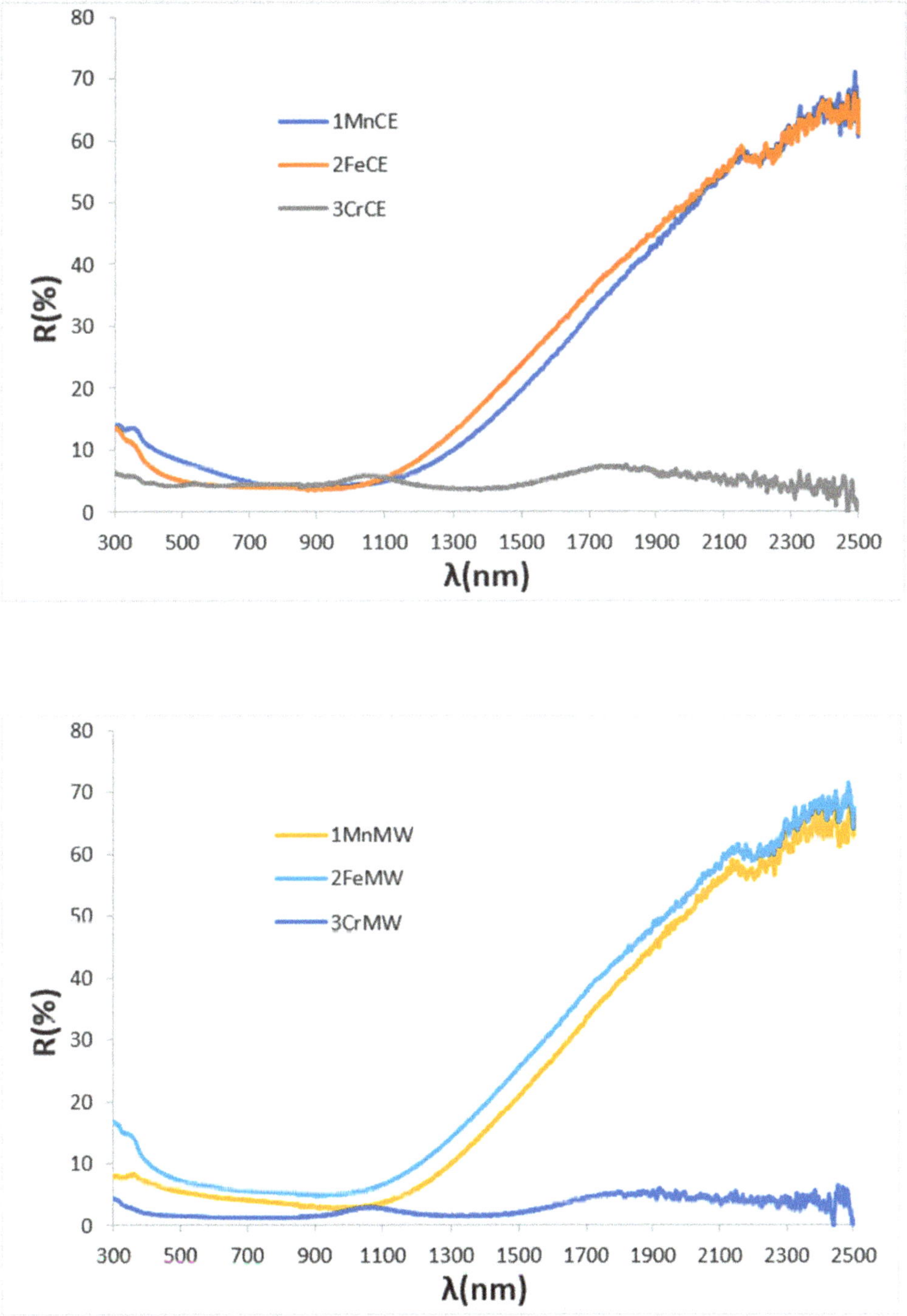

Figure 8. UV–Vis–NIR reflectance spectra of samples glazed using electric kiln.

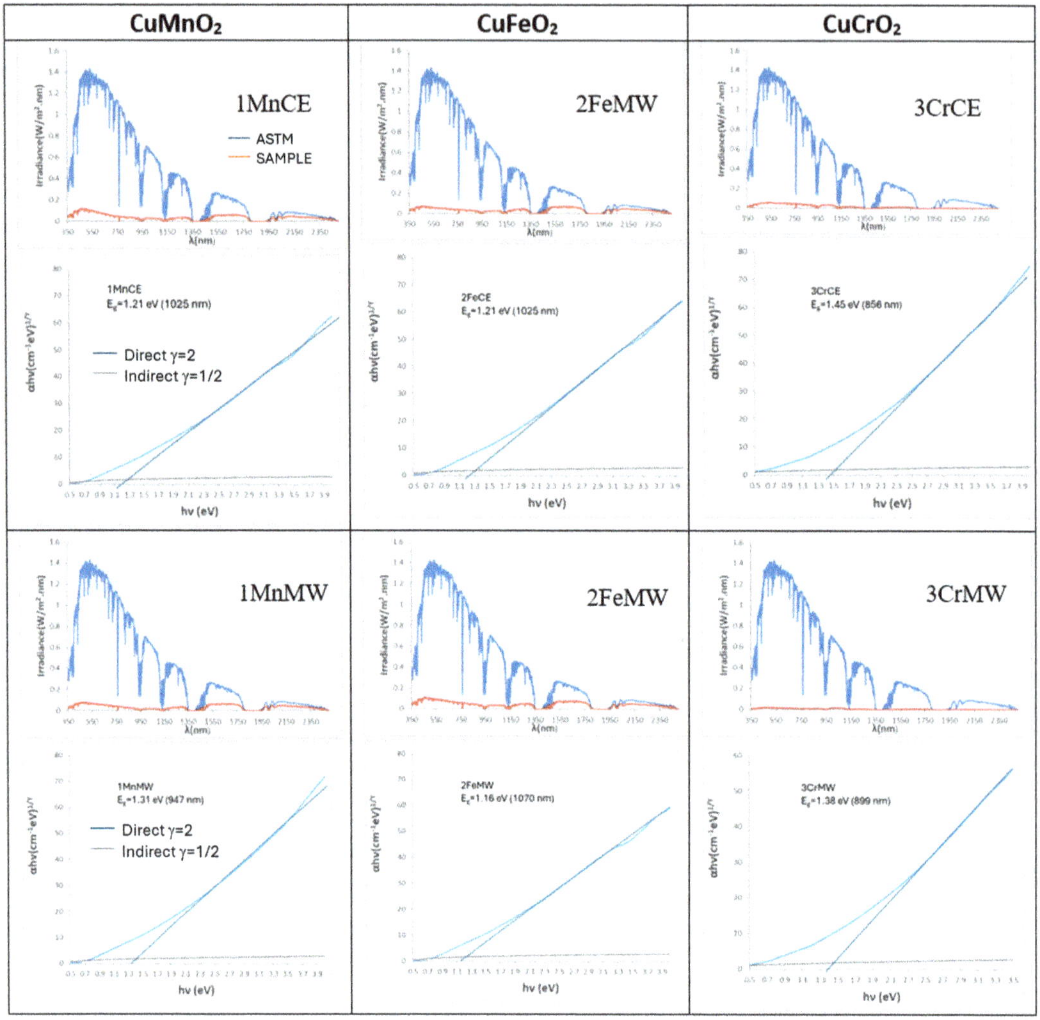

Figure 9. Tauc plots and solar spectra of glazed samples using electric kiln.

All microwave-fired glazes exhibit a dark black appearance with pinhole defects, with the most pronounced in the manganese sample and also present in 2FeMW and 3CrCE, alongside a greenish undertone ($a^* \leq 0$). These features are attributed to the behaviour of Cu^{2+} ions within the glass: octahedrally coordinated Cu^{2+} produces a green hue, while square-planar coordination can yield blue coloration [44]. In the manganese and iron samples, the presence of Cu^{2+} compounds ($CuMn_2O_4$ and CuO) and the aggressive interaction with the molten glass promote the partial exsolution and reduction of Cu^{2+} to Cu^+, releasing oxygen and resulting in the formation of the observed pinholes Equation (3) [44,45]:

$$2CuO_{(glaze/green\ shade)} \rightarrow Cu_2O_{(glaze,\ red\ shade)} + \tfrac{1}{2} O_2 \quad (3)$$

Crystalline phase stability dictates Cu^{2+} behaviour and optical properties: stable phases, as in mcconnellite, limit Cu^{2+} reduction and maintain blackness, whereas decomposed phases, such as crednerite and delafossite, promote Cu^{2+} reduction, resulting in greenish undertones and pinholes.

In microwave-glazed samples from electrically fired powders, R_{Vis} decreases in the order 1MnCE < 2CrCE < 3FeCE, corresponding to a blackness trend of 1MnCE > 2CrCE > 3FeCE, with higher ΔE* values linked to the greenish hue (b*) (Table 1).

For microwave-glazed samples, R_{Vis} follows the sequence 2FeMW > 1MnMW > 2CrMW, consistent with decreasing ΔE* and bandgap values. Mcconnellite exhibits an intense black appearance (R_{Vis} = 2.63, ΔE* = 4.3) and a reduced bandgap relative to the corresponding powder. Compared to electrically fired counterparts, microwave-glazed Mn and Cr samples show lower R_{Vis}, indicating a deeper black shade, whereas Fe displays higher R_{Vis}. UV–Vis–NIR spectra (Figure 10) reveal generally flat responses for microwave-glazed powders, with the Fe sample showing notable NIR reflectance. These findings suggest that microwave glazing promotes enhanced Cu exsolution from pigment particles while preserving their chemical stability better than electric kiln glazing.

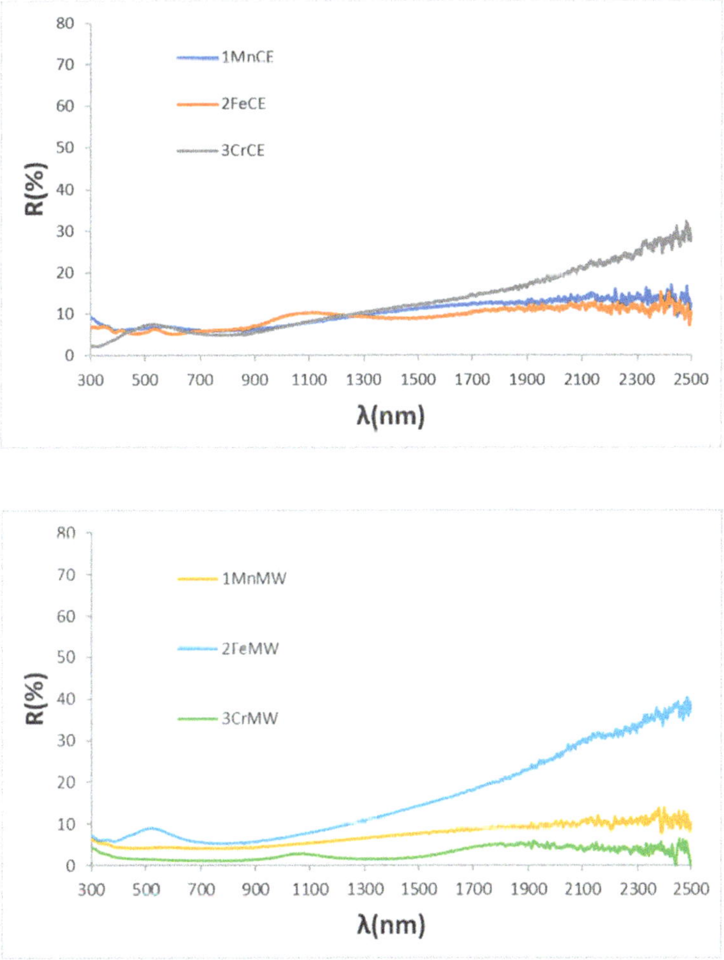

Figure 10. UV–Vis–NIR reflectance spectra of glazed samples using microwaves kiln.

Figure 11 shows the solar spectra and Tauc plots of microwave-glazed samples. Bandgap values vary between powders fired electrically and by microwave, reflecting differences in crystallization within the glass matrix. The spectra indicate strong solar absorption across the visible and near-infrared ranges, with sample 3CrMW exhibiting particularly high absorption, consistent with its deep-black colour (R_{Vis} = 2.63), slightly higher than its electric kiln counterpart (R_{Vis} = 1.41).

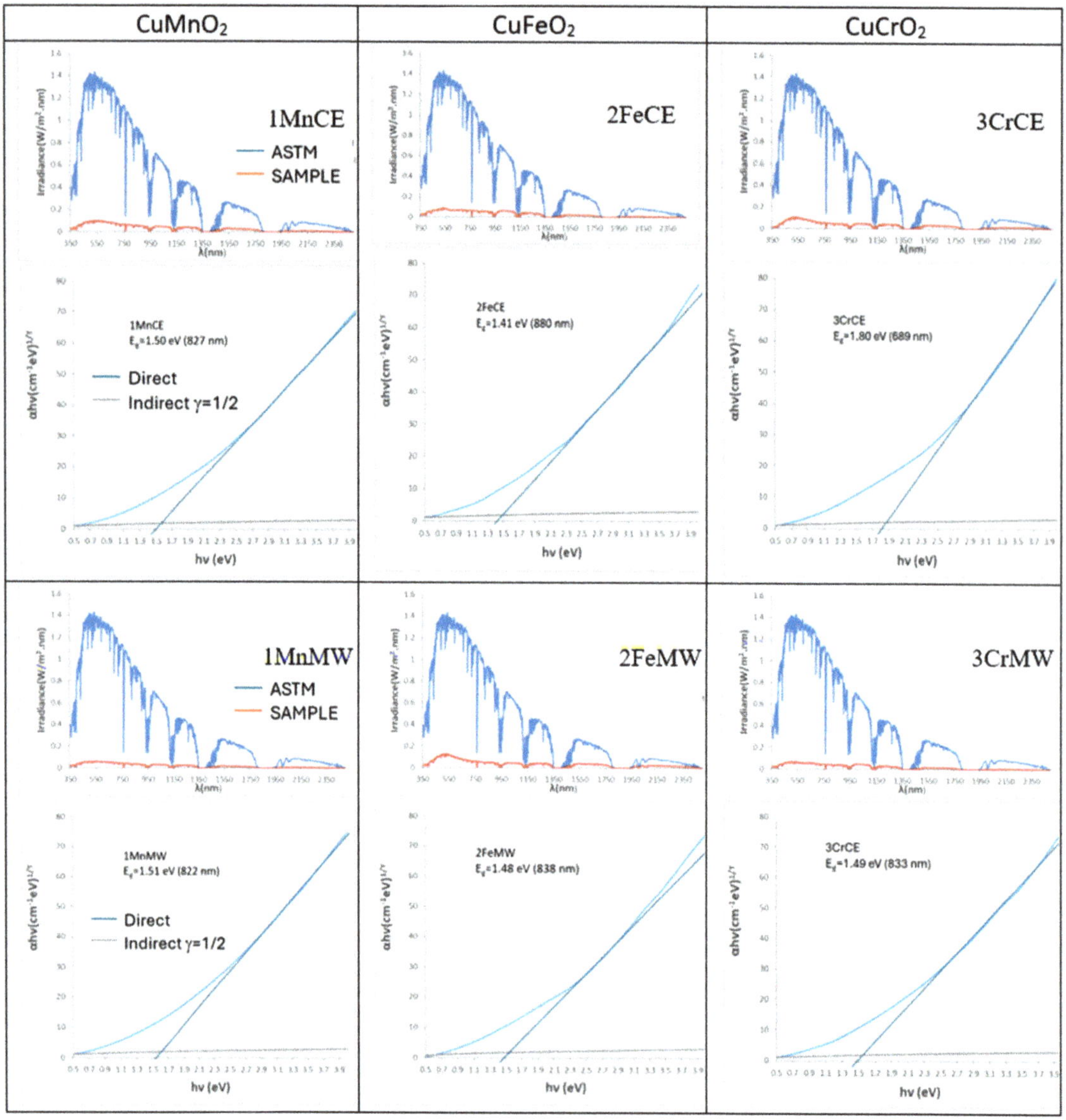

Figure 11. Tauc plots and solar spectra of glazed samples using microwaves kiln.

3.4. Effect of Temperature and Microwaves Power in Blackness of Mcconnellite

Table 5 summarizes the properties of mcconnellite ($CuCrO_2$) powders fired in an electric kiln at 1000 and 1100 °C for three hours and in a microwave kiln at 800 and 900 W for 30 min.

Table 5. Comparison of mcconnellite obtained with literature results (reference [17]).

SAMPLE	L*a*b*	$R_{Vis}/R_{NIR}/R$ (%)	E_g (eV)	ΔE*
POWDER				
Electric kiln				
3CrCE **1000 °C**/3 h	42.6/−5.4/0.1	13.8/14.1/13.9	1.26	23.1
3CrCE **1100 °C**/3 h	28.66/−4.17/1.17	6.10/8.26/7.03	1.40	9.5
Microwave Kiln				
3CrMW **800 W** 30 min	36.5/−4.8/−0.8	9.9/11.2/10.5	1.30	17
3CrMW **900 W** 30 min	31.05/−2.39/−0.15	6.98/7.73/7.31	1.40	11.1
GLAZED ELECTRIC KILN 5 wt% with glaze 1000				
Electric kiln				
3CrCE **1000 °C**/3 h	26.9/−0.1/−2.3	5.3/5.0/5.2	1.0	7.2
3CrCE **1100 °C**/3 h	24.70/−0.77/0.32	4.40/4.81/4.58	1.45	4.6
Microwave Kiln				
3CrMW **800 W** 30 min	11.3/−0.6/−3.0	1.4/1.9/1.6	0.93	9.5
3CrMW **900 W** 30 min	23.52/−0.02/−0.98	1.41/2.14/1.77	1.38	3.5

Mcconnellite is the only phase detected at both 1000 and 1100 °C, as shown in Figure 2 and reference [17]. However, the blackness of the powders increases (R_{Vis} decreases from 13.8 to 6.10), in agreement with the decrease in the green shade (b*), the bandgap, and the colour deviation ΔE* relative to Carbon Black.

Mcconnellite is the only crystalline phase detected at a microwave power of 800 W [17]. When the power is increased from 800 W to 900 W, the blackness of the powders increases (R_{Vis} decreases from 9.9 to 6.98). The weak diffraction peaks of spinel $CuCr_2O_4$ observed in this sample (Figure 2) indicate the presence of a "composite effect," which alters the reflection, absorption, and scattering of light, thereby producing distinct colour responses and enhanced blackness.

Glazing in a 1000 °C double firing frit increases blackness for electric furnace powders (R_{Vis} 5.30 to 4.40), whereas microwave-fired powders show nearly constant or slightly higher R_{Vis} (1.40 to 1.41), likely reflecting the persistence of the spinel phase and the loss of composite scattering effects when particles are embedded in the glassy matrix.

3.5. *Characterization of Mcconellite Doped with Iron and Manganese*

To assess firing temperature, glaze compatibility, and the effects of Fe and Mn doping, three samples were prepared: mcconnellite ($CuCrO_2$, reference), Fe-doped $(Cu_{0.8}Fe_{0.2})CrO_2$, and Fe–Mn co-doped $(Cu_{0.6}Mn_{0.2}Fe_{0.2})CrO_2$. All were fired in an electric kiln at 1200 °C for three hours.

As shown in Figure 12, the fired powders appear black. Mcconnellite exhibits a slight greenish undertone (b* = −1.2), while the doped samples are nearly neutral. The undoped sample shows higher visible reflectance (R_{Vis} = 8.89%) than the doped powders ($R_{Vis} \approx 3\%$), corresponding to a deeper black colour.

Figure 13 shows the XRD patterns of the synthesized powders. The mcconnellite sample contains only the $CuCrO_2$ phase. In the Fe-doped sample $(Cu_{0.8}Fe_{0.2})CrO_2$, mcconnellite remains predominant, with additional medium-intensity peaks corresponding to spinel $Cu(Cr,Fe)_2O_4$. In the Fe–Mn co-doped sample, the spinel $Cu(Cr,Fe,Mn)_2O_4$ phase dominates, accompanied by residual mcconnellite reflections. Spinels, known black pigments [14,15], contribute to increased light backscattering (composite effect), resulting in enhanced colour strength. Accordingly, the doped powders exhibit lower R_{Vis} values than pure mcconnellite.

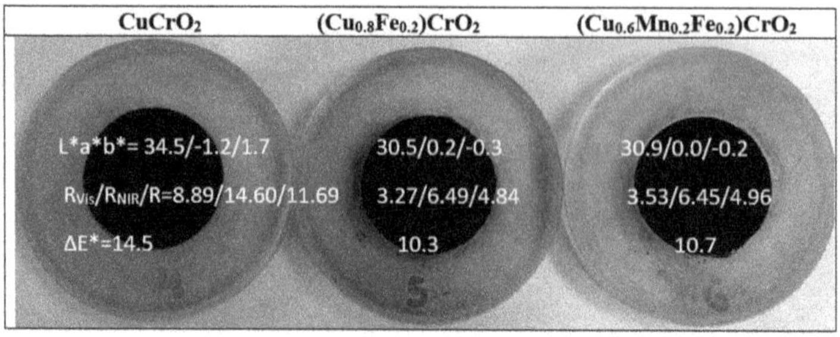

Figure 12. Powders fired at 1200 °C for 3 h of $CuCrO_2$, Fe-doped $(Cu_{0.8}Fe_{0.2})CrO_2$, and Fe- and Mn-doped $(Cu_{0.6}Mn_{0.2}Fe_{0.2})CrO_2$.

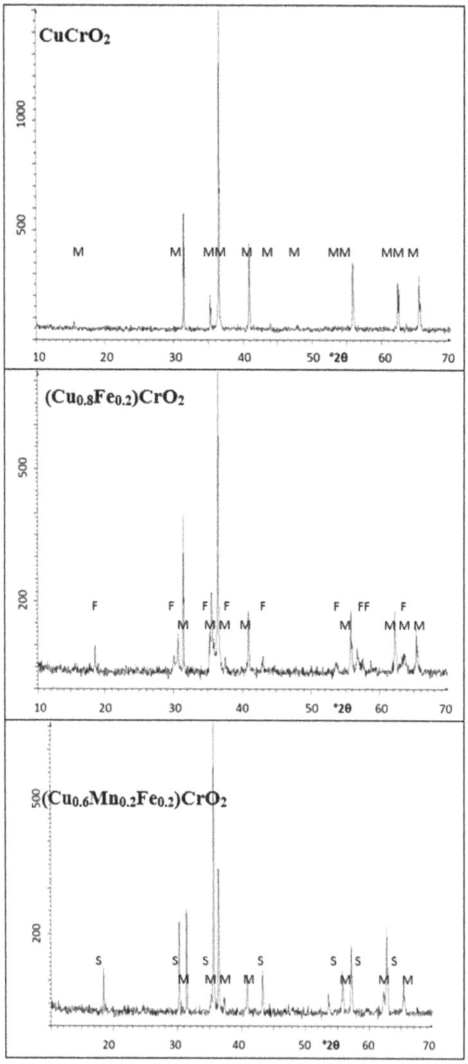

Figure 13. XRD of powders $CuCrO_2$, Fe-doped $(Cu_{0.8}Fe_{0.2})CrO_2$, and Fe- and Mn-doped $(Cu_{0.6}Mn_{0.2}Fe_{0.2})CrO_2$. Crystalline phases: M (mcconnellite $CuCrO_2$), F (spinel $Cu(Cr,Fe)_2O_4$), S (spinel $Cu(Cr,Fe,Mn)_2O_4$).

Figure 14 presents the UV–Vis–NIR diffuse reflectance and the corresponding solar absorption spectra of powders. The mcconnellite shows higher reflectivity, consistent with its lower blackness. Characteristic absorption minima at 430, 620, and 1200 nm (Cr^{3+}, octahedral coordination) and at 680 nm (Cu^{+}) are clearly visible. In the doped samples, overlapping bands from Mn^{3+} and Fe^{3+} hinder individual band assignment. The solar absorption spectra confirm higher overall absorbance for the doped compositions across the solar range.

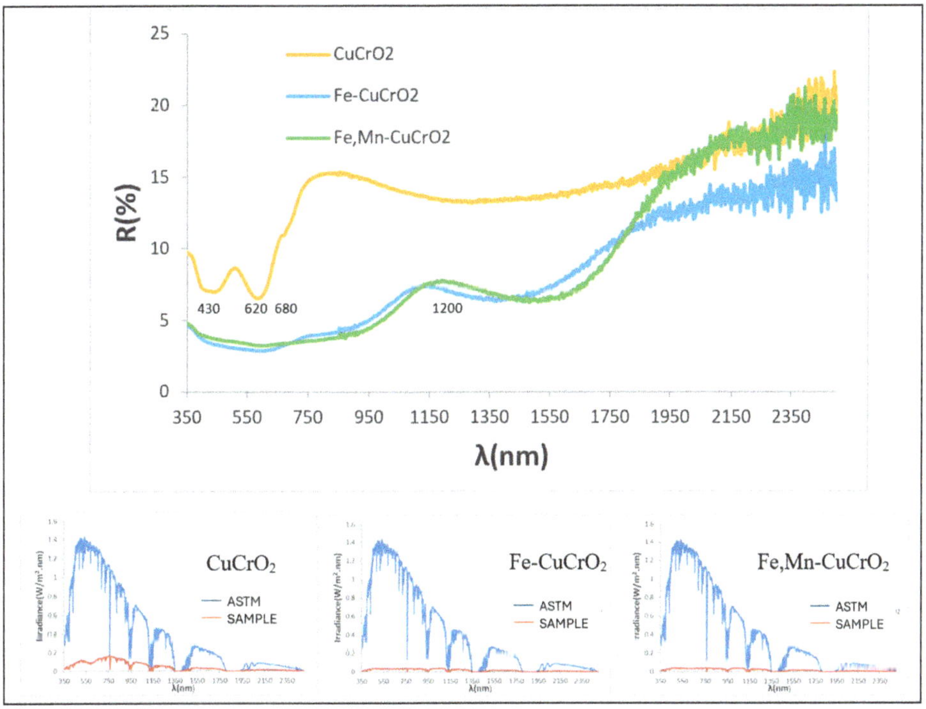

Figure 14. UV–Vis–NIR diffuse reflectance spectra of powders $CuCrO_2$, Fe-doped $(Cu_{0.8}Fe_{0.2})CrO_2$ and Fe- and Mn-doped $(Cu_{0.6}Mn_{0.2}Fe_{0.2})CrO_2$ and its respective solar spectra absorption.

Figure 15 shows images of 5 wt% glazed samples in a porcelain glaze (maturation point 1190 °C; composition in wt%: SiO_2 67, K_2O 3, CaO 12.5, MgO 1.5, ZnO 6, Al_2O_3 10; average grain size of 7–30 μm, supplied by Torrecid S.A. l'Alcora, Spain), including their colour and reflectance characteristics. The mcconnellite sample, having lost the composite effect observed in the powder form due to particle dispersion within the glassy matrix, now exhibits the deepest black shade with a slight greenish undertone (R_{Vis} = 4.37, b* = −2.5). In contrast, the doped samples display a visibly brown hue, with a* values of 4.9 and 3.1 for the Fe-doped and Fe–Mn co-doped compositions, respectively. When the pigments are glazed, the attack of this aggressive glassy matrix can dissolve and decompose the pigment particles, leading to the crystallization of highly stable spinels such as zinc spinels (ZnO should therefore be avoided in glazes to preserve pigment integrity) [44,45]. The high-temperature porcelain glaze supplied by Torrecid S.A. contains approximately 6 wt% ZnO; consequently, multicomponent brown spinels with a general composition of $(Zn,Fe,Mn,Cu)(Fe,Mn,Cr)_2O_4$ may form (see brown spinels in the CPMA classification [26]). These phases are difficult to analyze in glassy matrices containing dispersed particles.

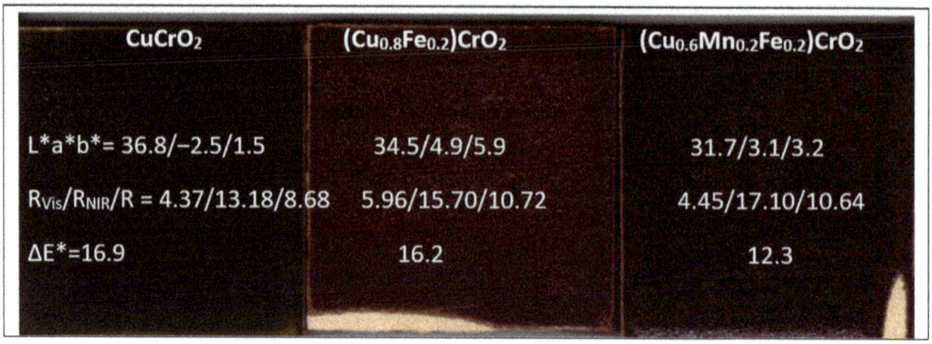

Figure 15. Images of 5 wt% glazed samples in glaze for porcelain (1190 °C) and alkyd paint (weight ratio paint/pigment/water = 7:2:5) of $CuCrO_2$, Fe-doped $(Cu_{0.8}Fe_{0.2})CrO_2$, and Fe- and Mn-doped $(Cu_{0.6}Mn_{0.2}Fe_{0.2})CrO_2$ and its respective solar spectra absorption.

Figure 16 shows the UV–Vis–NIR diffuse reflectance spectra of 5 wt% pigmented glazes on porcelain (fired at 1190 °C) and their corresponding solar absorption spectra. The three samples exhibit similar spectral profiles; however, the mcconnellite-glazed sample shows slightly lower reflectance in the yellow–red wavelength region (550–700 nm), resulting in its visually black appearance, in contrast to the brown hues of the doped glazes. In all cases, reflectance in the NIR region increases, indicating that these glazed surfaces are promising candidates for selective solar absorber applications. Among them, mcconnellite demonstrates the highest absorption in the visible range [47].

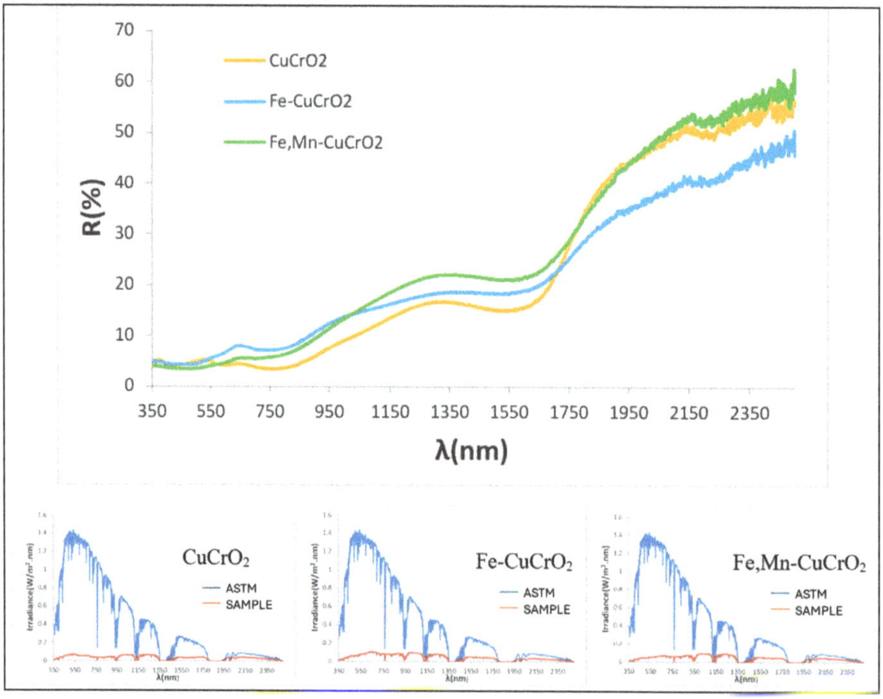

Figure 16. UV–Vis–NIR diffuse reflectance spectra of 5 wt% glazed samples in glaze for porcelain (1190 °C) $CuCrO_2$, Fe-doped $(Cu_{0.8}Fe_{0.2})CrO_2$, and Fe- and Mn-doped $(Cu_{0.6}Mn_{0.2}Fe_{0.2})CrO_2$ and its respective solar spectra absorption.

4. Conclusions

Cuprate delafossite phases, such as $CuMnO_2$ (crednerite) or $CuFeO_2$, as well as iron- and manganese-doped mcconnellite composites, were analyzed as candidates for producing intense black ceramic pigments using conventional solid-state synthesis, employing both electric kiln and fast dielectric (microwave) firing, and compared with mcconnellite ($CuCrO_2$) used as a reference.

In the case of crednerite, the electrically fired powder exhibits a composite of spinel and crednerite, whereas the microwave-fired powder consists almost exclusively of spinel, resulting in slightly lower blackness. When glazed in an electric kiln, spinel $CuMn_2O_4$ is detected in the microwave-glazed sample, while no diffraction peaks or crystallization halos are observed for the electrically fired sample. No evidence of crednerite preservation after interaction with the molten glaze is found. Therefore, the highly stable spinel $CuMn_2O_4$ is identified as the crystalline phase responsible for the black coloration in this system. Glazed samples prepared in a microwave kiln show identical behaviour.

For delafossite, both the electrically and microwave-fired powders contain a delafossite–tenorite ($CuFeO_2$–CuO) composite, with weak tenorite peaks in the electrically fired sample. The so-called "composite effect" accounts for the slightly higher blackness of the tenorite-rich microwave-fired powder (R_{Vis} = 6.45% versus 6.87% for the electrically fired sample). After glazing in an electric kiln, conventional XRD patterns display broad crystallization halos together with a diffraction peak at 39° 2θ, corresponding to the intense [114] plane reflection of tenorite, indicating the decomposition of delafossite within the molten glass. Similar behaviour is observed for the samples glazed using a microwave kiln.

For mcconnellite, a single-phase material is obtained in the electrically fired powder, whereas the microwave-fired sample shows residual oxide peaks. The higher blackness of mcconnellite explains the slightly lower R_{Vis} value for the electrically fired sample (R_{Vis} = 6.10% versus 6.98% for the microwave-fired sample). Upon glazing in an electric kiln, both grazing-incidence and conventional XRD analyses confirm the crystallization of $CuCrO_2$ in all cases. The persistence of mcconnellite in the molten glass demonstrates its high thermal stability and excellent performance as a black ceramic pigment. Comparable results are obtained for samples glazed in a microwave kiln.

Using a microwave kiln, the blackness of the samples improves compared to electric firing. Among the powders, crednerite stands out with R_{Vis} = 4.85%, corresponding to 95.15% absorption of visible light. In contrast, for the glazed samples, although an improvement in blackness was observed in all cases (except for the mcconnellite sample) when using the microwave kiln, the optimal process remains glazing in the electric kiln. This is because glazing with the microwave kiln leads to numerous pinhole defects, likely associated with copper reduction promoted by microwave processing.

For samples glazed in an electric kiln, crednerite exhibited an R_{Vis} value of 5.10 (94.9% absorption), whereas mcconnellite showed the highest blackness, with R_{Vis} = 1.41 (98.59% absorption). Among the samples glazed in a microwave kiln, crednerite reached R_{Vis} = 4.27 (95.73% absorption), and mcconnellite achieved R_{Vis} = 2.63 (97.37% absorption), which was slightly inferior to the corresponding glazed in the electric furnace.

An increase in the electric firing temperature from 1000 to 1100 °C resulted in an enhancement of mcconnellite blackness. In contrast, increasing the microwave power from 800 W to 900 W produced very similar values (R_{Vis} ≈ 1.40%).

The thermal stability of mcconnellite's blackness was confirmed at 1200 °C, together with its compatibility with porcelain glazes with high maturation temperature (1190 °C). The partial substitution of copper by iron and manganese further improved colour performance. The undoped mcconnellite powder maintained an intense black colour at 1200 °C (R_{Vis} = 8.89%), although this value was higher than those recorded for the doped composites (R_{Vis} = 3.27 and 3.23 for the Fe-doped and Fe–Mn-doped samples, respectively). The pigment also exhibited good resistance to high-maturation porcelain glazes (1190 °C), with R_{Vis} = 4.37, which was slightly inferior to the doped counterparts (R_{Vis} = 5.96 and 4.45 for the Fe- and Fe–Mn-doped samples, respectively), which displayed a brownish coloration attributed to the loss of the composite effect.

Among the delafossites investigated, only mcconnellite subjected to microwave-assisted firing (R_{Vis} = 1.40%, corresponding to 98.60% visible light absorption) emerges as a promising candidate for an ultra-black ceramic pigment (R_{Vis} < 1%), following further optimization under ZnO-free glaze conditions and a firing temperature of approximately 1000 °C. Considering the blackness in powder form, microwave-fired crednerite (R_{Vis} = 4.85; 95.15% absorption) and the Fe- and Fe–Mn-doped mcconnellite composites (R_{Vis} = 3.27 and 3.23%, respectively) appear as potential candidates for deep-black pigment applications (RVis < 3%), benefiting from the composite effect between the delafossite and chromium spinel phases. In addition, microwave-fired crednerite demonstrated promising deep-black behaviour in glazed samples (R_{Vis} = 4.27; 95.73% absorption).

Author Contributions: Conceptualization, G.M., G.M.-A. and M.L.; methodology, G.M. and C.D.; software, G.M.-A.; validation, G.M., C.D. and V.E.; formal analysis, G.M.-A.; investigation, C.D. and V.E.; resources, G.M.; data curation, C.D.; writing—original draft preparation, C.D., G.M. and G.M.-A.; writing—review and editing, G.M. and C.D.; visualization, V.E.; supervision, G.M. and M.L.; project administration, G.M. and M.L.; funding acquisition, G.M. and M.L. All authors have read and agreed to the published version of the manuscript.

Funding: This research was funded by Jaume I University (UJI) by UJI-B2021-73 and MEFD Spain by PID2024-158224NB-I00 funded by MCIN/AEI//10.13039/501100011033 and ERDF/EU.

Data Availability Statement: The original contributions presented in this study are included in the article. Further inquiries can be directed to the corresponding author.

Acknowledgments: The authors acknowledge Jaume I University (UJI) its financial support (UJI-B2021-73), the MEFD Spain and the technical support of the Servei Central d'Instrumentació Científica (SCIC) (UJI).

Conflicts of Interest: The authors declare no conflicts of interest.

Abbreviations

SSA	Selective solar absorber
IR	Infrared
XRD	X-ray Diffraction
SEM	Scanning electron microscopy with energy-dispersive spectroscopy
ICDD	International Centre for Diffraction Data
UV-Vis-NIR	Ultraviolet-visible-near-infrared

References

1. Friedel, C. Sur une combinaison naturelle des oxydes de fer et de cuivre, et sur la reproduction de l'atacamite. *Comptes Rendus De L'académie Des Ciences i* **1873**, *77*, 211–214.
2. Shannon, R.D.; Rogers, D.B.; Prewitt, C.T. Chemistry of noble metal oxides. I. Syntheses and properties of ABO2 delafossite compounds. *Inorg. Chem.* **1971**, *10*, 713–718. [CrossRef]
3. Tanaka, M.; Hasegawa, M.; Higuchi, T. Origin of the metallic conductivity in $PdCoO_2$ with delafossite structure. *Physica B* **1998**, *245*, 157–163. [CrossRef]
4. Marquardt, M.A.; Ashmore, N.A.; Cann, D.P. Crystal chemistry and electrical properties of the delafossite structure. *Thin Solid Film.* **2006**, *496*, 146–156. [CrossRef]
5. Kawazoe, H.; Yasukawa, M.; Hyodo, H.; Kurita, M.; Yanagi, H.; Hosono, H. P-type electrical conduction in transparent thin films of $CuAlO_2$. *Nature* **1997**, *389*, 939–942. [CrossRef]

6. Daoua, R.; Frésarda, R.; Eyerta, V.; Héberta, S.; Maignana, A. Unconventional aspects of electronic transport in delafossite oxides. *Sci. Technol. Adv. Mater.* **2017**, *18*, 919–938. [CrossRef]
7. Pabst, A. Crystal structure and density of delafossite. *Am. Mineral.* **1938**, *23*, 175–176.
8. Kohler, B.U.; Jansen, M. Synthesis and crystal structure of 2H-$CuAlO_2$. *Z. Krist.* **1983**, *165*, 313–314. [CrossRef]
9. Poienar, M.; Banica, R.; Sfirloaga, P.; Ianasi, C.; Mihali, C.V.; Vlazan, P. Microwave-assisted hydrothermal synthesis and catalytic activity study of crednerite-type $CuMnO_2$ materials. *Ceram. Int.* **2018**, *44*, 6157–6161. [CrossRef]
10. Jackson, J.J.; Puretzky, A.A.; More, K.L.; Rouleau, C.M.; Eres, G.; Geohegan, D.B. Pulsed growth of vertically aligned nanotube arrays with variable density. *ACS Nano* **2010**, *4*, 7573–7581. [CrossRef] [PubMed]
11. Monrós, G. Black Inorganic Pigments: From Black to Ultra-Black Behaviour, A Microstructural Approach. *J. Miner. Mater. Sci.* **2025**, *6*, 1119. [CrossRef]
12. Marques, V.; Krings, W.; Reis, J.; Gorb, S.; Guillermo-Ferreira, R. Ultrablack color in velvet ant cuticle. *Beilstein J. Nanotechnol.* **2024**, *15*, 1554–1565. [CrossRef] [PubMed]
13. Brenner, M.; Hearing, V.J. The protective role of melanin against UV damage in human skin. *Photochem. Photobiol.* **2008**, *84*, 539–549. [CrossRef] [PubMed]
14. Calbo, J.; Sorlí, S.; Llusar, M.; Tena, M.A.; Monrós, G. Minimisation of toxicity in nickel ferrite black pigment. *Br. Ceram. Trans.* **2004**, *103*, 3–9. [CrossRef]
15. Calbo, J.; Mestre, A.; García, A.; Tena, M.A.; Llusar, M.; Monrós, G. Multicomponent black coloured spinels from alkoxides. *J. Sol-Gel Sci. Technol.* **2003**, *26*, 191–194. [CrossRef]
16. Monrós, G.; Delgado, C.; Monrós-Andreu, G.; Badenes, J.; Llusar, M. Black Coloured Glazes with Tetragonal $CuCr_2O_4$ Ceramic Pigment as Selective Solar Absorbers for Integral Ceramic Solar Collectors. *J. Sol. Energy Res. Updates* **2024**, *11*, 78–92. [CrossRef]
17. Monrós, G.; Badenes, J.; Delgado, C.; Monrós-Andreu, G.; Llusar, M. New Mcconnellite Ceramic Pigment as a Selective Solar Absorber: Effects of Microwave Firing and Rare Earth Doping. *Materials* **2025**, *18*, 1520. [CrossRef]
18. Feng, G.; Liu, T.; Mu, J.; Jiang, F.; Guo, J.; Chiao, L.; Wu, Q.; Zhang, X.; Liu, J.; Liang, J. Effect of mineralizer on the synthesis of (Ni, Mn) co-doping $CuFe_5O_8$ pigment and its glazing performance. *Open Ceram.* **2025**, *23*, 100822. [CrossRef]
19. Monrós, G.; Cerro, S.; Badenes, J.A.; Llusar, M. Black Cool Pigments for Urban Heat Island (UHI) Control: From Cr-Hematite to Mn-Melilite. *J. Sol. Energy Res. Updates* **2021**, *8*, 27–44. [CrossRef]
20. Calbo, J.; Gargori, C.; Sorlí, S.; Badenes, J.; Tena, M.A.; Monrós, G. Síntesis de pigmentos cerámicos mediante radiación microondas. *Bol. Soc. Esp. Ceram. Vidr.* **2007**, *46*, 14–20. [CrossRef]
21. Blosi, M.; Dondi, M.; Albonetti, S.; Baldi, G.; Barzanti, A.; Zanelli, C. Microwave-assisted synthesis of Pr–$ZrSiO_4$, V–$ZrSiO_4$ and Cr–$YAlO_3$ ceramic pigments. *J. Eur. Cer. Soc.* **2009**, *29*, 2951–2957. [CrossRef]
22. Bhattacharya, M.; Basak, T. A review on the susceptor assisted microwave processing of materials. *Energy* **2016**, *97*, 306–338. [CrossRef]
23. CIE Comission International de l'Eclairage. *Recommendations on Uniform Color Spaces, Colour Difference Equations, Psychometrics Colour Terms*; Bureau Central de la CIE: Paris, France, 1978.
24. Tauc, J.; Grigorovici, R.; Vancu, A. Optical Properties and Electronic Structure of Amorphous Germanium. *Phys. Status Solidi* **1966**, *15*, 627–637. [CrossRef]
25. *ASTM G173-03*; Standard Tables for Reference Solar Spectral Irradiance at Air Mass 1.5: Direct Normal and Hemispherical for a 37 Degree Tilted Surface. American Society for Testing and Materials (ASTM) International: Conshohocken, PA, USA, 2020.
26. CPMA. *Classification and Chemical Description of the Complex Inorganic Color Pigments*, 4th ed.; Alexandria Dry Color Manufaturers Association: Alexandria, VA, USA, 2010.
27. Monrós, G. *Pigment, Ceramic, in Encyclopedia of Color Science and Technology*; Luo, R., Ed.; Springer: Berlin/Heidelberg, Germany, 2014; ISBN 978-1-4419-8070-0/978-3-642-27851-8. Available online: https://link.springer.com/rwe/10.1007/978-1-4419-8071-7_181 (accessed on 1 October 2025).
28. *ASTM D3265-21*; Standard Test Method for Carbon Black—Tint Strength. American Society for Testing and Materials (ASTM) International: West Conshohocken, PA, USA, 2021.
29. *ISO 5435:2017(E)*; Rubber Compounding Ingredients, Carbon Black, Determination of Tinting Strength. International Organization for Standardization (ISO): Geneva, Switzerland, 2017.
30. Ferch, H.; Seibold, K. Zur Bedeutung und Existenz von Primärteilchen bei hochdispersen Stoffen. *Farbe Lack* **1984**, *90*, 88–98.
31. Ross, C.; Nwaubani, S. Increasing Pigment Black Jetness with Color Pigment Blending, DCL Corporation. Available online: https://www.pigments.com/wp-content/uploads/RETEC-2020-Increasing-Pigment-Black-Jetness-with-Color-Pigment-Blending.pdf (accessed on 1 July 2025).
32. Driessens, F.C.M.; Rieck, G.D. Phase Equilibria in the System Cu-Mn-O. *Z. Anorg. Und Allg. Chem.* **1967**, *351*, 48–62. [CrossRef]
33. Trari, M.; Topferb, J.; Dordorc, P.; Grenierc, J.C.; Pouchardc, M.; Doumercc, J.P. Preparation and physical properties of the solid solutions Cu1+xMn1-xO2 (0<x<0.2). *J. Solid State Chem.* **2005**, *178*, 2751–2758. [CrossRef]

34. Huang, X.; Ni, C.; Zhao, G.; Irvine, J.T.S. Oxygen storage capacity and thermal stability of the $CuMnO_2$–CeO_2 composite system. *J. Mater. Chem. A* **2015**, *24*, 1–7. [CrossRef]
35. Bessekhouad, Y.; Gabes, Y.; Bouguelia, A.; Trari, M. The physical and photo electrochemical characterization of the crednerite $CuMnO_2$. *J. Mater. Sci.* **2007**, *42*, 6469–6476. [CrossRef]
36. Solsona, B.; Hutchings, G.J.; Garcia, T.; Taylor, S.H. Improvement of the catalytic performance of CuMnOx catalysts for CO oxidation by the addition of Au. *New J. Chem.* **2004**, *28*, 708–711. [CrossRef]
37. Shannon, R.D. Revised effective ionic radii and systematic studies of interatomic distances in halides and chalcogenides. *Acta Cryst.* **1976**, *A32*, 751–767. [CrossRef]
38. Huanga, Y.; Wangan, S.; Tsaib, A.; Kameoka, S. Reduction behaviors and catalytic properties for methanol steam reforming of Cu based spinel compounds CuX_2O_4 (X=Fe, Mn, Al, La). *Ceram. Int.* **2014**, *40*, 4541–4551. [CrossRef]
39. Qiu, X.; Liu, M.; Sunada, K.; Miyauchi, M.; Hashimoto, K. A facile one-step hydrothermal synthesis of rhombohedral $CuFeO_2$ crystals with antivirus property. *Chem. Commun.* **2012**, *48*, 7365–7367. [CrossRef]
40. Matasov, A.; Bush, A.; Kozlov, V.; Stash, A. 3R-$CuCrO_2$ delafosite: Crystal growth, crystal structure, dielectric and DC conductivity properties. *Discov. Mater.* **2024**, *4*, 73. [CrossRef]
41. Sanam, P.K.J.; Shah, M.; Pradyumnan, P.P. Structure induced modification on thermoelectric and optical properties by Mg doping in $CuCrO_2$ nanocrystals. *Solid State Commun.* **2022**, *353*, 114855. [CrossRef]
42. Dondi, M.; Zanelli, C.; Ardit, M.; Cruciani, G.; Mantovani, L.; Tribaudino, M.; Andreozzi, G.B. Ni-free, black ceramic pigments based on Co—Cr—Fe—Mn spinels: A reappraisal of crystal structure, colour and technological behaviour. *Ceram. Int.* **2013**, *39*, 9533–9547. [CrossRef]
43. Prasanna, T.R.S. Angle dependence of the integrated intensity in thin film X-ray diffraction. *arXiv* **2017**, arXiv:1612.01805v2. [CrossRef]
44. Verger, L.; Olivier, D.; Rousse, G.; Cotte, M.; Cormier, M.L. The Stability of Gahnite Doped with Chromium Pigments in Glazes from the French Manufacture of Sèvres. *J. Am. Ceram. Soc.* **2016**, *100*, 86–95. [CrossRef]
45. Monrós, G.; Badenes, J.A.; Llusar, M.I. Ecofriendly High NIR Reflectance Ceramic Pigments Based on Rare Earths Compared with Classical Chromophores Prepared by DPC Method. *Ceramics* **2022**, *5*, 614–641. [CrossRef]
46. Schorne-Pinto, J.; Cassayre, L.; Presmanes, L.; Barnabé, A. Insights on the Stability and Cationic Nonstoichiometry of $CuFeO_2$ Delafossite. *Inorg. Chem.* **2019**, *58*, 6431–6444. [CrossRef]
47. Li, Y.; Lin, C.; Huang, J.; Chi, C.; Huang, B. Spectrally Selective Absorbers/Emitters for Solar Steam Generation and Radiative Cooling-Enabled Atmospheric Water Harvesting. *Glob. Chall.* **2021**, *5*, 2000058. [CrossRef]

materials

MDPI

Article

Thermoelectric Properties of $NbCoNi_xSn$ (x = 0–1)

Moritz Thiem, Ruijuan Yan, Anke Weidenkaff and Wenjie Xie *

Materials and Resources, Department of Materials Science, Technical University of Darmstadt, 64287 Darmstadt, Germany; moritz.thiem@mr.tu-darmstadt.de (M.T.); kate66998@gmail.com (R.Y.); anke.weidenkaff@mr.tu-darmstadt.de (A.W.)
* Correspondence: wenjie.xie@mr.tu-darmstadt.de

Abstract

The half-Heusler (HH) compound NbCoSn, with 18 valence electrons, is a promising thermoelectric (TE) material due to its favourable electrical properties and excellent thermal and chemical stability. Enhancing its TE performance typically involves doping and microstructure engineering. In this study, Ni was introduced into NbCoSn to form $NbCoNi_xSn$ (x = 0–1), and the effects of Ni content on the microstructure and TE properties were systematically investigated. At low doping levels ($x \leq 0.05$), Ni occupies interstitial sites, forming $NbCoNi_xSn$ solid solutions. At higher concentrations ($x > 0.05$), full-Heusler (FH) secondary phases emerge, resulting in HH–FH composites. The introduction of Co/Ni interstitials enhances TE performance by creating in-gap electronic states and increasing phonon scattering through point defects. A clear structural transition from HH to FH phases is observed with increasing Ni content. The highest figure of merit, $ZT \approx 0.52$ at 975 K, was obtained for $NbCoNi_{0.05}Sn$, comparable to the best values reported for this system.

Keywords: thermoelectric materials; half-Heusler; NbCoSn; microstructural modification; secondary phase

1. Introduction

In our modern world, global climate change is the essential driver for research on sustainable energy. Global energy consumption growth leads to the need for new low-carbon energy sources [1,2]. Thermoelectric materials can directly convert waste heat into electricity and offer a way to improve the sustainability of already existing systems, such as combustion engines, sunlight, or by-products of other heat-producing processes [3–6]. Moreover, thermoelectrics offer various flexible applications such as temperature control, refrigeration, and power generation [7–9]. The dimensionless thermoelectric figure of merit $ZT = \sigma S^2/(\kappa_L + \kappa_e)$ (with σ being the electrical conductivity, S the Seebeck coefficient, and κ_e and κ_L the contributions of electrons and lattice to the total thermal conductivity κ_{tot}) is used to determine the effectiveness of the system.

The class of half-Heusler (HH) compounds have demonstrated promising thermoelectric performance, with peak figures of merit (ZT) reaching values as high as 1.5 [10–13]. A key advantage of these materials lies in their tuneable electronic properties, which can be adjusted through compositional modifications without compromising the integrity of their crystal structure [14,15]. Moreover, HH compounds are also attractive for practical applications due to their low toxicity, high mechanical robustness, and excellent thermal stability [16–19]. As a result, current research efforts are focused on further enhancing the thermoelectric efficiency of HH materials. This can be achieved either by increasing the power factor ($PF = S^2\sigma$) or by reducing their inherently high thermal conductivity

κ [19,20]. To this end, various approaches have been explored, including nanostructuring [17], the introduction of point defects [21], plastic deformation [22], and band structure engineering [23].

NbCoSn is a member of the half-Heusler (HH) compound family and has shown promising thermoelectric (TE) performance, with reported ZT values in the range of 0.5 to 0.6 [24–26]. It is a ternary half-Heusler compound that crystallises in the cubic MgAgAs-type structure (space group $F\overline{4}3m$, No. 216) [27,28]. NbCoSn is an intrinsic n-type semiconductor with an 18 valence electron count (VEC) and a bandgap of approximately 0.987 eV. This bandgap corresponds to roughly 10 $k_B T_0$, where k_B is the Boltzmann constant and T_0 is the application temperature, making NbCoSn a promising candidate for high-temperature thermoelectric applications [29]. So far, many studies have been carried out to improve the TE properties of NbCoSn compounds via tuning the chemical compositions, such as Sb substitution on the Sn site [24], Pt substitution on the Co site [25], and Ta substitution on the Nb site [26].

Among all the studies, it was noticed that excess Co is always observed in directly melted NbCoSn samples, synthesised by high-temperature techniques (arc-melting or induction-melting), despite being designed as the nominal Nb:Co:Sn = 1:1:1 composition [16,25,30–32], indicating that there are some stable intrinsic interstitial Co defects. This excess in Co content originates from the formation of secondary phases, mainly Nb_3Sn and Sn, during the melting procedure and cannot be fully eliminated. It has been reported that interstitial Co defects can introduce in-gap states between the conduction and valence bands, thereby significantly enhancing electrical conductivity [33,34]. Recently, Yan et al. [16] demonstrated that in $NbCo_{1-x}Ni_xSn$ compounds, Ni atoms not only substituted on the 4c lattice position of Co but also occupied the vacant 4d position in the half-Heusler lattice, leading to the partial formation of the full-Heusler (FH) phase in the half-Heusler matrix (Figure 1). The occupation of the 4d site with Ni was verified by theoretical calculations, which also showed that interstitial Ni defects can create in-gap states that can improve the thermoelectric performance. Although these point defects can enhance phonon scattering, the reported thermal conductivities were still higher than comparable literature values [25], limiting the overall improvement in ZT.

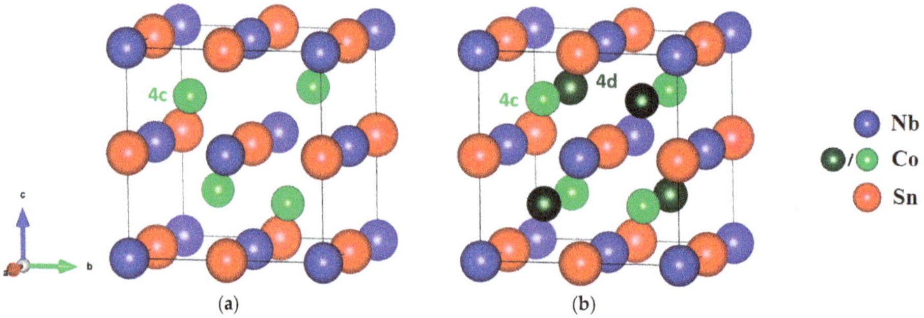

Figure 1. Crystal structure of (**a**) HH NbCoSn compound and (**b**) FH $NbCo_2Sn$ compound.

To further enhance the ZT and understand the role of Ni in the system, $NbCoNi_xSn$ (x = 0–1) samples are prepared by arc-melting, followed by additional ball milling and spark plasma sintering. Laboratory X-ray Diffraction (XRD) and Scanning Electron Microscopy (SEM) were employed to show the change in sample composition from a pure NbCoSn HH phase to a NbCoNiSn FH phase with negligible secondary phases. Due to the introduction of Ni, electrical properties are enhanced at low doping levels, leading to a maximum ZT of ~0.52 at 975 K. Moreover, the introduction of ball milling to the $NbCoNi_xSn$

systems proves to be a valuable addition to the synthesis since thermal conductivities were reduced drastically.

2. Experimental Methods

$NbCoNi_xSn$ (with nominal compositions x = 0, 0.1, 0.05, 0.1, 0.15, 0.2, 0.25, 0.3, 0.4, 0.5, and 1.0) synthesis starts by arc-melting the stoichiometric amounts of Nb (wire, 99.999%), Co (bulk, 99.999%), Ni (wire, 99.99%), and Sn (shot, 99.99%) under an inert argon atmosphere. To ensure a homogeneous melting process, the ingots are flipped, crushed, and re-melted several times. Afterwards, the bulk samples were ground into powder using high-energy ball milling (Pulverisette 7, Fritsch, Idar-Oberstein, Germany) for 2 h. The powders were loaded in a Ø20 mm graphite die and spark plasma sintered at 1223 K for 5 min. Subsequently, the cylinder-shaped samples are cut into discs and bars for phase characterisation, structural analysis, and TE property measurements.

The powder X-ray diffraction (XRD) patterns were all measured using Mo Kα1 radiation in a STOE STAD diffractometer (Darmstadt, Germany). For analysis of phase composition and phase distribution, Scanning Electron Microscopy is used (SEM, TESCAN, VEGA 3, Dortmund, Germany) with an energy-dispersive X-ray spectrometer (EDX, EDAX Genesesis, AMTEK Gmbh, Unterschleissheim, Germany). For high-temperature measurements of the Seebeck coefficient S and electrical resistivity σ, a ZEM-3 device (ULVAC-RIKO, Yokohama, Japan) was used. The charge-carrier concentration n_H and mobility μ_H were measured at room temperature with an applied magnetic field of −520 to 520 mT by a commercial Hall measurement system HT-Hall (Fraunhofer IPM, Freiburg, Germany). Measurements of the thermal conductivity were performed based on the formula $\kappa = DdC_p$, with D being the thermal diffusivity, d being the densities of the samples, and C_p being the specific heat. D and d were, respectively, measured by laser flash analysis (LFA 457, NETZSCH, Selb, Germany) and an Archimedes kit. C_p was determined using the Dulong–Petit law. Uncertainties for characterisation techniques for thermoelectric properties are ±3% for σ, ±5% for S, and ±10% for κ.

3. Results and Discussion

3.1. Phase Composition and Microstructure Characterisation

The laboratory XRD patterns of nominal $NbCoNi_xSn$ samples (x = 0, 0.05, 0.1, 0.15, 0.2, 0.25, 0.3, 0.4, 0.5, and 1; hereafter referred to as Ni-x for simplicity) are shown in Figure 2. The main phases are consistent with the MgAgAs-type HH structure and the $AlCu_2Mn$-type FH structure. Typical secondary phases for these compounds are Nb_3Sn, Co_7Nb_6, and elemental Sn [24,35]. Hardly any secondary-phase patterns were observed in the XRD shown in Figure 2, with intensities too low to provide precise information on their phase identity and quantity. The intensities of secondary phases (Nb_3Sn, Co_7Nb_6, and Sn) do not enhance with Ni contents, indicating the secondary phases in all Ni-x samples are more or less the same. The main change in XRD patterns is the increase in intensity of the FH-phase peaks with increasing Ni content. Previous studies on $NbCo_{1-x}Ni_xSn$ systems demonstrated that Ni occupying the vacant 4d position is energetically favourable. Furthermore, the introduction of Ni can stabilise the FH phase ($NbCo_{2-x}Ni_xSn$), and it is more stable than the $NbCo_2Sn$ phase [16]. Therefore, the increase in the FH phase in the $NbCoNi_xSn$ system was expected as the quantity of the FH phase is directly related to the amount of excess Ni introduced into the system.

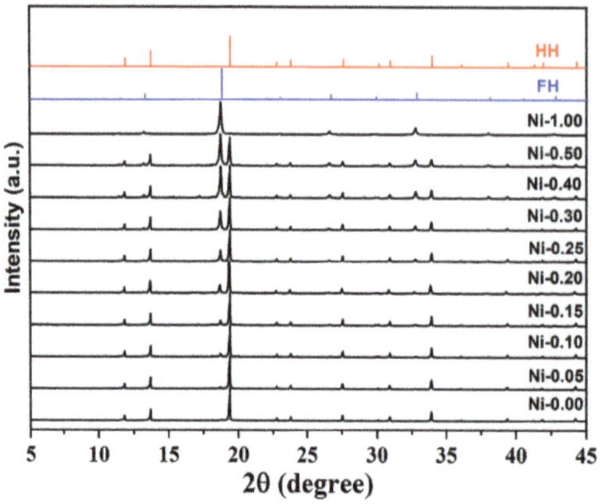

Figure 2. Powder XRD pattern of $NbCoNi_xSn$ patterns after SPS with literature peaks of half-Heusler (HH) and full-Heusler (FH) phases.

The linear increase in the FH phase with increasing Ni content is also reflected in the measured densities (Table 1). Overall, the measured densities were close to the theoretical densities. For the theoretical density calculations, it was assumed that the nominal excess Ni occupies the 4d lattice site. The good agreement between measured and theoretical densities supports the validity of this assumption.

Table 1. Measured densities compared to the theoretical density after SPS of $NbCoNi_xSn$ compounds.

Sample	Measured Density [g/cm^3]	Theoretical Density [g/cm^3]	Percentage (%)
Ni-0	8.510	8.522	99.86
Ni-0.05	8.537	8.566	99.67
Ni-0.1	8.516	8.610	98.91
Ni-0.15	8.645	8.654	99.90
Ni-0.2	8.668	8.698	99.66
Ni-0.25	8.757	8.742	100.17
Ni-0.3	8.785	8.786	99.99
Ni-0.4	8.828	8.874	99.48
Ni-0.5	8.943	8.962	99.79
Ni-1	9.258	9.401	98.48

To further investigate phase distribution and microstructure in the Ni-x compounds, back-scattered electron (BSE) imaging and corresponding energy-dispersive X-ray (EDX) mapping of Ni were conducted on polished sample surfaces, and typical results are presented in Figure 3. The EDX analysis results for the two main phases are summarised in Table 2. An excess of Co was detected in nearly all phases. This excess can be attributed to the formation of secondary phases such as Nb_3Sn and elemental Sn, which result from the high-temperature melting synthesis. These secondary phases are difficult to eliminate using conventional melting methods and contribute to an excess of Co in the remaining Heusler phases, even when the nominal composition is adjusted accordingly [16,25,30–32]. Regarding the effect of adding Ni into the NbCoSn system, the volume fraction of the FH phase increases with increasing Ni content and is homogeneously distributed within the

HH matrix. EDX Ni mapping confirms that the FH phase is the primary host for Ni, while point analyses reveal only trace amounts of Ni in the HH phase. Fracture surface SEM imaging was also carried out to estimate grain sizes (Supporting Information Figure S1), which were found to be below 10 μm for all samples. This is approximately an order of magnitude smaller than those reported in earlier studies on NbCoNiSn compounds [16].

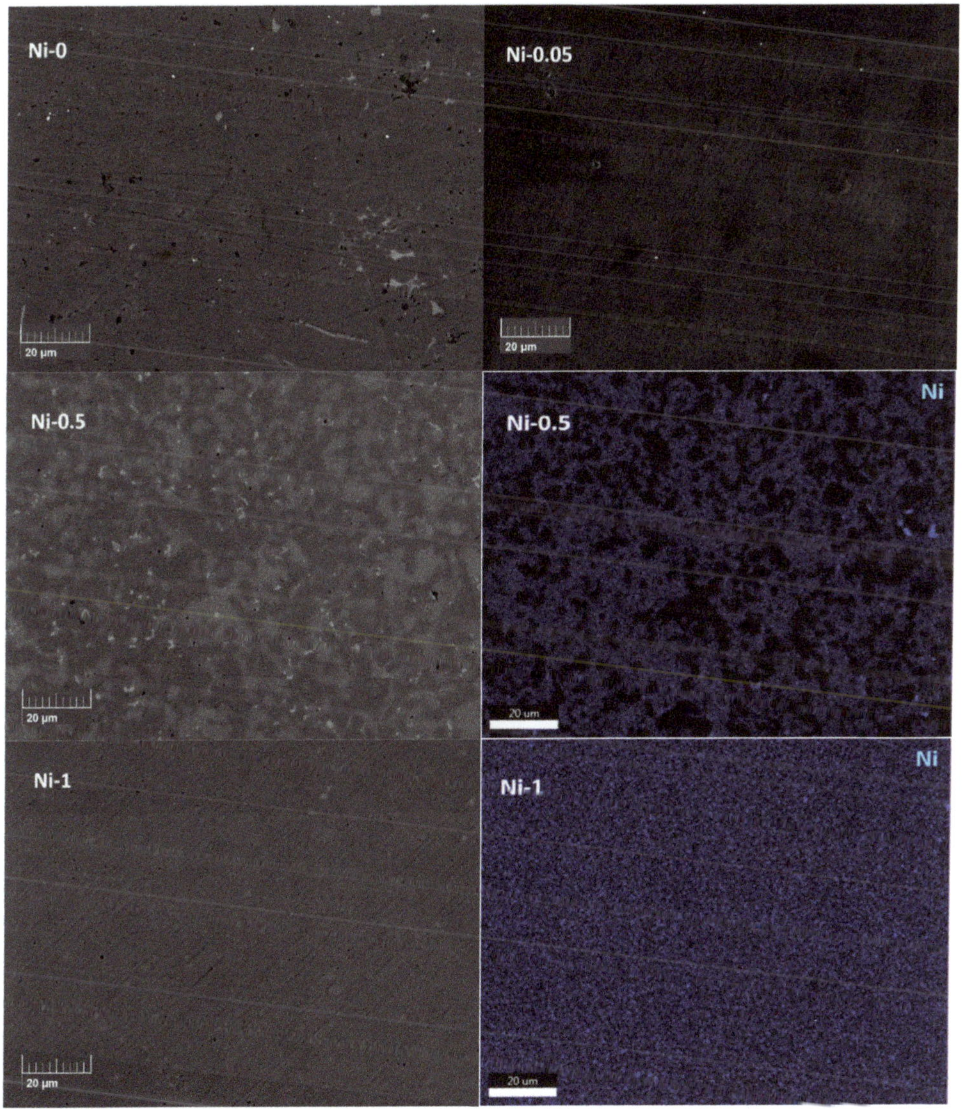

Figure 3. BSE image and corresponding EDX elemental mapping of Ni of representative $NbCoNi_xSn$ compounds.

Table 2. Actual and nominal phase compositions of $NbCoNi_xSn$ samples after SPS, measured with EDX.

Nominal Composition	Actual Phase Composition Light Phase	Actual Phase Composition Dark Phase
NbCoSn	$Nb_{0.89}Co_{1.12}Sn$	
$NbCoNi_{0.05}Sn$	$Nb_{0.89}Co_{1.12}Ni_{0.03}Sn$	$Nb_{0.90}Co_{1.12}Ni_{0.04}Sn$
$NbCoNi_{0.10}Sn$	$Nb_{0.89}Co_{1.12}Ni_{0.12}Sn$	$Nb_{0.71}Co_{1.12}Ni_{0.69}Sn$
$NbCoNi_{0.15}Sn$	$Nb_{0.93}Co_{1.04}Ni_{0.10}Sn$	$Nb_{0.73}Co_{1.35}Ni_{0.51}Sn$
$NbCoNi_{0.20}Sn$	$Nb_{0.90}Co_{1.10}Ni_{0.15}Sn$	$Nb_{0.82}Co_{1.67}Ni_{0.47}Sn$
$NbCoNi_{0.25}Sn$	$Nb_{0.90}Co_{0.97}Ni_{0.10}Sn$	$Nb_{0.75}Co_{1.30}Ni_{0.71}Sn$
$NbCoNi_{0.30}Sn$	$Nb_{0.89}Co_{1.04}Ni_{0.12}Sn$	$Nb_{0.74}Co_{1.60}Ni_{0.61}Sn$
$NbCoNi_{0.40}Sn$	$Nb_{0.87}Co_{0.98}Ni_{0.13}Sn$	$Nb_{0.72}Co_{1.37}Ni_{0.74}Sn$
$NbCoNi_{0.50}Sn$	$Nb_{0.87}Co_{0.99}Ni_{0.13}Sn$	$Nb_{0.79}Co_{1.35}Ni_{0.80}Sn$
$NbCoNi_{1.00}Sn$	$Nb_{0.91}Co_{1.10}Ni_{1.10}Sn$	

3.2. Transport Properties

The results of the thermoelectric property measurements are displayed in Figure 4. For all samples, electrical conductivity decreases with increasing temperature, indicating a metal-like transport behaviour. At room temperature, the electrical conductivity increases almost linearly with Ni content from $4.69 \cdot 10^4$ S/m for Ni-0 to $53.41 \cdot 10^4$ S/m for Ni-1. Conversely, the Seebeck coefficient decreases with increasing Ni content. This inverse relationship between the electrical conductivity and Seebeck coefficient suggests an increase in charge-carrier concentration with higher Ni content, consistent with the expected interdependence of these properties. To verify this, Hall effect measurements were performed, and the results are shown in Figure 5. Hall measurements (Figure 5) confirm the assumption of an increase in charge-carrier concentration with higher Ni content. The observed trends in the electrical conductivity and Seebeck coefficient can be attributed to the evolving phase composition, specifically the increasing proportion of the full-Heusler phase relative to the half-Heusler phase. The FH phase has an intrinsically low Seebeck coefficient (sample Ni-1, Figure 4b) but a higher electrical conductivity, and vice versa for the HH phase. This interpretation is consistent with the microstructural analysis from SEM, which revealed a progressive increase in FH phase content with increasing Ni addition.

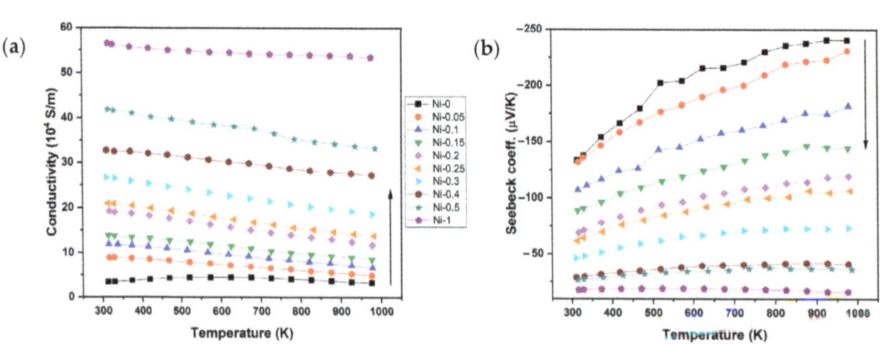

Figure 4. Temperature dependence measurements of (**a**) electrical conductivity and (**b**) Seebeck coefficient.

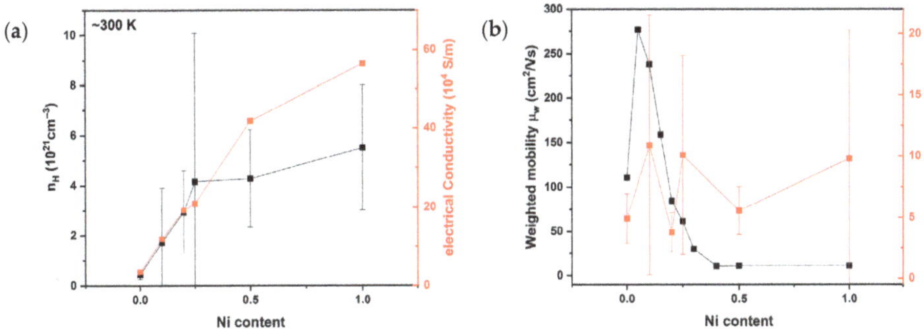

Figure 5. Dependences of (**a**) the charge-carrier concentration and electrical conductivity, and (**b**) weighted mobility and Hall mobility of the Ni content in $NbCoNi_xSn$ compound at room temperature.

The observed increase in charge-carrier concentration in this system can arise through several mechanisms. First, the substitution of Co with Ni introduces an additional electron per Ni atom. Second, excess Ni can occupy interstitial 4d sites, forming in-gap states that reduce the band gap and thereby increase carrier concentration—a phenomenon previously confirmed in other HH systems such as MNiSn (M = Ti, Zr, Hf) [36,37]. Third, the formation of the half-metallic full-Heusler (FH) phase also contributes free carriers to the system [38].

Among these, direct substitution of Co by Ni is considered energetically less favourable than interstitial occupation of the 4d site [16], and can therefore be excluded as a dominant mechanism. Given the continued increase in electrical conductivity and decrease in Seebeck coefficient with increasing FH phase content, the FH phase is identified as the primary contributor to the increased charge-carrier concentration at higher Ni contents. Nevertheless, the incorporation of Ni in the HH phase via 4d site occupation also plays a role, especially at lower Ni concentrations. Based on structural analysis, the solubility limit of Ni interstitials appears to lie between $x = 0.05$ and 0.1, beyond which the FH phase begins to form.

The band structure modification induced by Ni occupation on the 4d site is also reflected in the increase in the charge-carrier effective mass (m^*), as shown in Figure 6. The effective mass of sample Ni-0 without any doping showed an effective mass of about $3.7m_e$, which is significantly lower than the values reported in the literature for NbCoSn compounds of around $5.4m_e$ to $6.5m_e$ [16,24,25]. However, measurement errors from the Hall and Seebeck coefficient measurements are large enough to assume that the presented effective mass is close to the literature values. The samples with high Ni doping content ($0.05 < x < 0.5$) possess a larger effective mass, which is due to the doping and FH secondary phase [39]. For the samples with x = 0.5 and x = 1, the SPB model may not be valid.

As the charge-carrier concentration increases, a corresponding decrease in mobility is typically expected. This trend is also observed in the present study through the calculated weighted mobility ($_{w}$), as shown in Figure 5b. The weighted mobility was determined using Equation (1) and is preferred over Hall mobility for further analysis, particularly in low-mobility materials where it provides a more reliable representation of carrier transport characteristics [40]. The equation is as follows:

$$\mu_W = \frac{3h^3\sigma}{8\pi e(2m_e k_B T)^{3/2}}\left[\frac{\exp\left(\frac{|S|}{k_B/e}-2\right)}{1+\exp\left[-5\left(\frac{|S|}{k_B/e}-1\right)\right]}+\frac{\frac{3}{\pi^2}\frac{|S|}{k_B/e}}{1+\exp\left[5\left(\frac{|S|}{k_B/e}-1\right)\right]}\right] \quad (1)$$

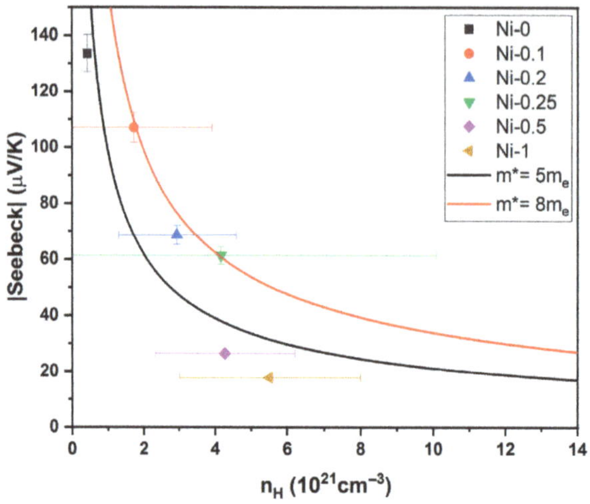

Figure 6. Seebeck coefficient over the Hall carrier concentration for $NbCoNi_xSn$ samples at room temperature.

The weighted mobility initially increases with the addition of Ni but begins to decline as the Ni content is further increased. The initial rise in $_w$ can be attributed to a significant increase in electrical conductivity, while the Seebeck coefficient does not decrease proportionally (Figure 4). This behaviour is consistent with the observed power factor, which is directly influenced by both the electrical conductivity and Seebeck coefficient. As expected from the trend in weighted mobility, the power factor reaches its maximum at low Ni concentrations and then decreases with further Ni additions. The measured power factor values are presented in Figure 7.

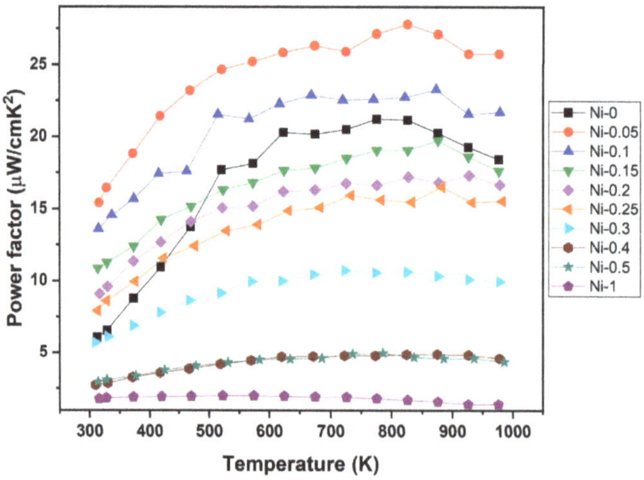

Figure 7. Temperature dependence of the power factor for $NbCoNi_xSn$ samples (Ni-*x*).

The power factors of all samples increase with temperature and reach a plateau at higher temperatures, primarily due to the rising absolute value of the Seebeck coefficient which stabilises at around T = 800 K. This behaviour aligns with trends reported in earlier studies on NbCoSn-based compounds [24,25]. A maximum power factor was achieved for the Ni-0.05 sample at 825 K with ~2.77 mW m^{-1} K^{-2}. Samples with Ni content up to 0.1 exhibit a higher power factor compared to undoped NbCoSn. However, with further increasing Ni content, power factors decrease almost linearly. This can be attributed to the gradual transition from HH to the FH phase, since the FH NbCoNiSn phase exhibits a much lower Seebeck coefficient and is therefore suboptimal for TE applications. For a higher power factor the Ni content in the range of 0 to 0.1 would have to be finely tuned.

3.3. Thermal Transport Properties

The change in total thermal conductivity of $NbCoNi_xSn$ is shown in Figure 8a. Compared with undoped NbCoSn (Ni-0), Ni-0.05 possesses the lowest thermal conductivity, and the thermal conductivity of the other Ni-*x*´s steadily increases with higher Ni contents. Moreover, the trend changes, with thermal conductivity increasing with increasing temperature for FH compounds and decreasing with higher temperature for HH compounds. The minimum thermal conductivity achieved was 4.48 W/mK for sample Ni-0.05 at T = 1023 K. Results of thermal conductivities for low Ni contents are comparable to the thermal conductivities of previous studies on NbCoSn compounds, ranging between 5 and 10 W m^{-1} K^{-1} [16,24,25,30,31,35,41]. The change in the trend of thermal conductivities is coherent with previously discussed results. Moreover, in comparison to a synthesis procedure without a ball milling step, the total thermal conductivity was reduced while maintaining a high power factor [16].

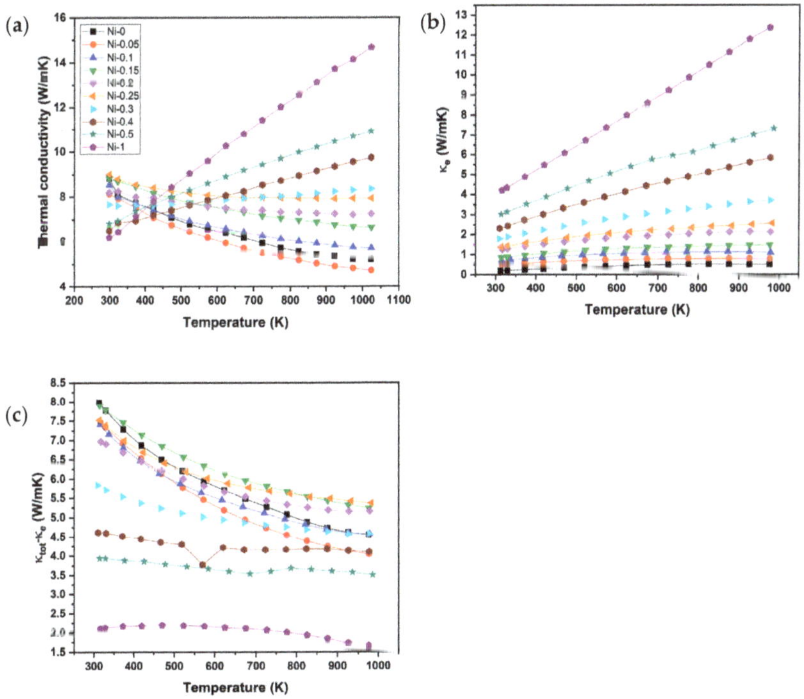

Figure 8. Temperature dependence of (**a**) total thermal conductivity, (**b**) electrical thermal conductivity, and (**c**) bipolar and lattice contribution to the total thermal conductivity of $NbCoNi_xSn$ compounds.

Figure 8b shows the electrical contribution to the thermal conductivity ($\kappa_e = L\sigma T$, with L being the Lorenz number) and the resulting bipolar and lattice thermal contributions. The Lorenz number was obtained according to Equation (2) as follows [42]:

$$L = 1.5 + \exp\left(-\frac{|S|}{116}\right) \quad (2)$$

With the calculated Lorenz numbers the electrical contribution to the thermal conductivity increases with increased Ni, consistent with the increase in charge-carrier concentration. In contrast, the combination of bipolar and lattice thermal conductivities does not follow the same linear trends as the measurements before. The lattice and bipolar thermal conductivity first decrease with added Ni up to Ni-0.1 and then increase and decrease again with the additional formation of the FH phase. The first decrease is attributed to enhanced phonon scattering caused by Ni-induced point defects in the HH phase, which suppress the lattice thermal conductivity. The subsequent variation is explained by the increasing presence of the FH phase, which alters phonon transport and affects both lattice and bipolar thermal contributions.

3.4. Figure of Merit

Due to simultaneous improvements of *PF* and reductions in κ, the TE performance ($ZT = S^2\sigma T/\kappa$) (Figure 9) of the NbCoSn sample was increased for low Ni additions <0.1. A maximum *ZT* value of ~0.52 was achieved at *T* = 975 K for the Ni-0.05 sample, which is an increase of about 48% compared to the pristine NbCoSn sample. Previous studies on NbCoSn compounds indicated a linear increasing trend of *ZT* throughout the measured temperature range [16,24,25,31]. This is only observed for sample Ni-0.05. For all other samples, the TE performance reaches a plateau at temperatures ~850 K without indication of further increases with higher temperatures. Overall, thermoelectric properties could only be enhanced for samples with Ni < 0.1, since an increase in *PF* with a simultaneously small decrease in thermal conductivity was only achieved in these samples. With additional Ni and therefore more of the FH phase, *PF* and thermal conductivity become less suitable for TE application.

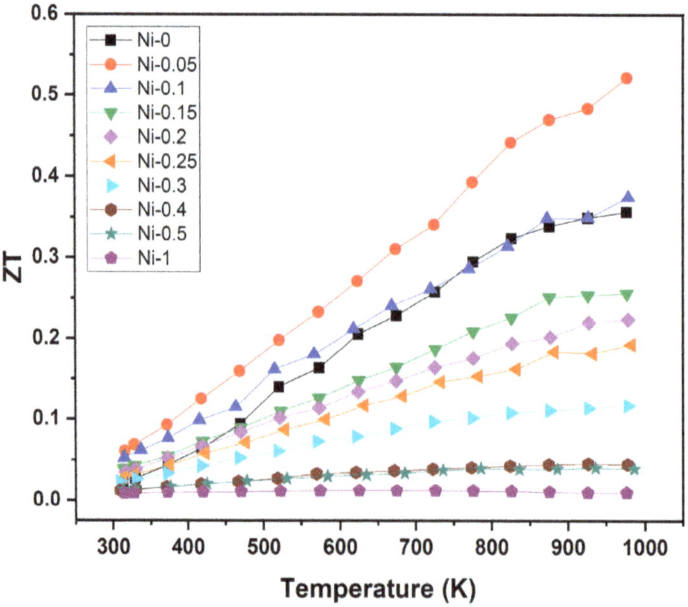

Figure 9. Figure of merit *ZT* of $NbCoNi_xSn$ compounds over temperature.

4. Conclusions

In summary, NbCoSn-based alloys with varying additional Ni content were successfully prepared via arc-melting and high-energy planetary ball milling, followed by SPS and further annealing. The structure and chemical properties of the synthesised samples were systematically characterised and analysed. Ni incorporation was found to increase thermoelectric properties in NbCoSn systems by simultaneously optimising electrical and thermal transport properties. Through controlled Ni addition, occupation of the 4d site was achieved, enabling a gradual transformation from the half-Heusler (HH) to the full-Heusler (FH) phase. This phase evolution was accompanied by corresponding changes in the microstructure and thermoelectric properties. Overall, a maximum power factor of ~2.77 $mWm^{-1}K^{-2}$ is obtained at 825 K. Furthermore, the lattice thermal conductivity is decreased due to the enhanced point defect scattering of phonons, leading to a peak *ZT* value of 0.52 at T = 975 K in the $NbCoNi_{0.05}Sn$ sample.

Supplementary Materials: The following supporting information can be downloaded at: https://www.mdpi.com/article/10.3390/ma18133189/s1, Figure S1: SEM pictures of fracture surfaces to compare grain sizes of $NbCoNi_xSn$ compounds. Figure S2: BSE Images (left side) and corresponding Ni elemental EDX mapping (right side).

Author Contributions: Conceptualization, W.X.; Methodology, M.T. and R.Y.; Validation, M.T.; Formal analysis, R.Y.; Investigation, M.T. and R.Y.; Resources, A.W.; Data curation, M.T.; Writing—original draft, M.T.; Writing—review & editing, A.W. and W.X.; Supervision, R.Y. and W.X.; Project administration, W.X.; Funding acquisition, A.W. and W.X. All authors have read and agreed to the published version of the manuscript.

Funding: This research received no external funding

Institutional Review Board Statement: Not applicable.

Informed Consent Statement: Not applicable.

Data Availability Statement: The original contributions presented in the study are included in the article/supplementary material, further inquiries can be directed to the corresponding author.

Acknowledgments: We would like to acknowledge Oliver Gutfleisch of TUDa for allowing us to use the SEM in his group.

Conflicts of Interest: The authors declare no conflicts of interest.

References

1. Houghton, J. Global warming. *Rep. Prog. Phys.* **2005**, *68*, 1343. [CrossRef]
2. Jackson, R.B.; Le Quéré, C.; Andrew, R.M.; Canadell, J.G.; Korsbakken, J.I.; Liu, Z.; Peters, G.P.; Zheng, B. Global energy growth is outpacing decarbonization. *Environ. Res. Lett.* **2018**, *13*, 120401. [CrossRef]
3. Burnete, N.V.; Mariasiu, F.; Depcik, C.; Barabas, I.; Moldovanu, D. Review of thermoelectric generation for internal combustion engine waste heat recovery. *Prog. Energy Combust. Sci.* **2022**, *91*, 101009. [CrossRef]
4. Mirhosseini, M.; Rezania, A.; Rosendahl, L. Power optimization and economic evaluation of thermoelectric waste heat recovery system around a rotary cement kiln. *J. Clean. Prod.* **2019**, *232*, 1321–1334. [CrossRef]
5. Sootsman, J.R.; Chung, D.Y.; Kanatzidis, M.G. New and old concepts in thermoelectric materials. *Angew. Chem. Int. Ed.* **2009**, *48*, 8616–8639. [CrossRef] [PubMed]
6. Xu, S.; Hong, M.; Shi, X.; Li, M.; Sun, Q.; Chen, Q.; Dargusch, M.; Zou, J.; Chen, Z.-G. Computation-guided design of high-performance flexible thermoelectric modules for sunlight-to-electricity conversion. *Energy Environ. Sci.* **2020**, *13*, 3480–3488. [CrossRef]
7. He, R.; Schierning, G.; Nielsch, K. Thermoelectric devices: A review of devices, architectures, and contact optimization. *Adv. Mater. Technol.* **2018**, *3*, 1870016. [CrossRef]
8. Russ, B.; Glaudell, A.; Urban, J.J.; Chabinyc, M.L.; Segalman, R.A. Organic thermoelectric materials for energy harvesting and temperature control. *Nat. Rev. Mater.* **2016**, *1*, 16050. [CrossRef]

9. Xu, Y.; Li, Z.; Wang, J.; Zhang, M.; Jia, M.; Wang, Q. Man-portable cooling garment with cold liquid circulation based on thermoelectric refrigeration. *Appl. Therm. Eng.* **2022**, *200*, 117730. [CrossRef]
10. Pedersen, S.V.; Croteau, J.R.; Kempf, N.; Zhang, Y.; Butt, D.P.; Jaques, B.J. Novel synthesis and processing effects on the figure of merit for NbCoSn, NbFeSb, and ZrNiSn based half-Heusler thermoelectrics. *J. Solid State Chem.* **2020**, *285*, 121203. [CrossRef]
11. Rogl, G.; Sauerschnig, P.; Rykavets, Z.; Romaka, V.V.; Heinrich, P.; Hinterleitner, B.; Grytsiv, A.; Bauer, E.; Rogl, P. (V,Nb)-doped half Heusler alloys based on {Ti,Zr,Hf}NiSn with high ZT. *Acta Mater.* **2017**, *131*, 336–348. [CrossRef]
12. Zhu, H.; He, R.; Mao, J.; Zhu, Q.; Li, C.; Sun, J.; Ren, W.; Wang, Y.; Liu, Z.; Tang, Z.; et al. Discovery of ZrCoBi based half Heuslers with high thermoelectric conversion efficiency. *Nat. Commun.* **2018**, *9*, 2497. [CrossRef] [PubMed]
13. Zhu, T.; Fu, C.; Xie, H.; Liu, Y.; Zhao, X. High efficiency half-Heusler thermoelectric materials for energy harvesting. *Adv. Energy Mater.* **2015**, *5*, 1500588. [CrossRef]
14. Klaer, P.; Kallmayer, M.; Blum, C.G.F.; Graf, T.; Barth, J.; Balke, B.; Fecher, G.H.; Felser, C.; Elmers, H.J. Tailoring the electronic structure of half-metallic Heusler alloys. *Phys. Rev. B* **2009**, *80*, 144405. [CrossRef]
15. Klaer, P.; Bos, T.; Kallmayer, M.; Blum, C.G.F.; Graf, T.; Barth, J.; Balke, B.; Fecher, G.H.; Felser, C.; Elmers, H.J. Charge transfer and tunable minority band gap at the Fermi energy of a quaternary $Co_2(Mn_xTi_{1-x})Ge$ Heusler alloy. *Phys. Rev. B* **2010**, *82*, 104410. [CrossRef]
16. Poon, S. Recent advances in thermoelectric performance of half-Heusler compounds. *Metals* **2018**, *8*, 989. [CrossRef]
17. Sakurada, S.; Shutoh, N. Effect of Ti substitution on the thermoelectric properties of (Zr,Hf)NiSn half-Heusler compounds. *Appl. Phys. Lett.* **2005**, *86*, 82105. [CrossRef]
18. Xie, W.; Weidenkaff, A.; Tang, X.; Zhang, Q.; Poon, J.; Tritt, T.M. Recent advances in nanostructured thermoelectric half-Heusler compounds. *Nanomaterials* **2012**, *2*, 379–412. [CrossRef]
19. Yan, R.; Xie, R.; Xie, W.; Shen, C.; Li, W.; Balke, B.; Yoon, S.; Zhang, H.; Weidenkaff, A. Effects of doping Ni on the microstructures and thermoelectric properties of Co-excessive NbCoSn half-Heusler compounds. *ACS Appl. Mater. Interfaces* **2021**, *13*, 34533–34542. [CrossRef]
20. Bos, J.-W.G.; Downie, R.A. Half-Heusler thermoelectrics: A complex class of materials. *J. Phys. Condens. Matter* **2014**, *26*, 433201. [CrossRef]
21. Yang, J.; Meisner, G.P.; Chen, L. Strain field fluctuation effects on lattice thermal conductivity of ZrNiSn-based thermoelectric compounds. *Appl. Phys. Lett.* **2004**, *85*, 1140–1142. [CrossRef]
22. Ghosh, S.; Shankar, G.; Karati, A.; Rogl, G.; Rogl, P.; Bauer, E.; Murty, B.S.; Suwas, S.; Mallik, R.C. Thermoelectric properties enhancement of $Ba_{0.2}Co_4Sb_{12}$ through dispersion of GaSb inclusions. *Phys. B Condens. Matter* **2021**, *606*, 412440. [CrossRef]
23. Fu, C.; Zhu, T.; Liu, Y.; Xie, H.; Zhao, X. Band engineering of high performance p-type FeNbSb based half-Heusler thermoelectric materials for figure of merit zT > 1. *Energy Environ. Sci.* **2015**, *8*, 216–220. [CrossRef]
24. He, R.; Huang, L.; Wang, Y.; Samsonidze, G.; Kozinsky, B.; Zhang, Q.; Ren, Z. Enhanced thermoelectric properties of n-type NbCoSn half-Heusler by improving phase purity. *APL Mater.* **2016**, *4*, 104804. [CrossRef]
25. Serrano-Sánchez, F.; Luo, T.; Yun, J.; Xie, W.; Le, C.; Auffermann, G.; Weidenkaff, A.; Zhu, T.; Zhao, X.; Alonso, J.A.; et al. Thermoelectric properties of n-type half-Heusler NbCoSn with heavy-element Pt substitution. *J. Mater. Chem. A* **2020**, *8*, 14822–14828. [CrossRef]
26. Wang, Q.; Huang, J.; Wang, C.; Luo, P.; Li, Z.; Liu, R.; Ma, Q.; Luo, J. Enhancement of thermoelectric properties in n-type NbCoSn half-Heusler compounds via Ta alloying. *ACS Appl. Energy Mater.* **2021**, *4*, 12458–12465. [CrossRef]
27. Graf, T.; Felser, C.; Parkin, S.S. Simple rules for the understanding of Heusler compounds. *Prog. Solid State Chem.* **2011**, *39*, 1–50. [CrossRef]
28. Jeitschko, W. Transition metal stannides with MgAgAs and $MnCu_2Al$ type structure. *Metall. Mater. Trans. B* **1970**, *1*, 3159–3162. [CrossRef]
29. Xi, L.; Yang, J.; Wu, L.; Yang, J.; Zhang, W. Band engineering and rational design of high-performance thermoelectric materials by first-principles. *J. Materiomics* **2016**, *2*, 114–130. [CrossRef]
30. Kimura, Y.; Tamura, Y.; Kita, T. Thermoelectric properties of directionally solidified half-Heusler compound NbCoSn alloys. *Appl. Phys. Lett.* **2008**, *92*, 12105. [CrossRef]
31. Li, S.; Zhu, H.; Mao, J.; Feng, Z.; Li, X.; Chen, C.; Cao, F.; Liu, X.; Singh, D.J.; Ren, Z.; et al. n-Type TaCoSn-based half-Heuslers as promising thermoelectric materials. *ACS Appl. Mater. Interfaces* **2019**, *11*, 41321–41329. [CrossRef] [PubMed]
32. Yan, R.; Xie, W.; Balke, B.; Chen, G.; Weidenkaff, A. Realizing p-type NbCoSn half-Heusler compounds with enhanced thermoelectric performance via Sc substitution. *Sci. Technol. Adv. Mater.* **2020**, *21*, 122–130. [CrossRef]
33. Chauhan, N.S.; Bathula, S.; Gahtori, B.; Mahanti, S.D.; Bhattacharya, A.; Vishwakarma, A.; Bhardwaj, R.; Singh, V.N.; Dhar, A. Compositional tailoring for realizing high thermoelectric performance in hafnium-free n-type ZrNiSn half-Heusler alloys. *ACS Appl. Mater. Interfaces* **2019**, *11*, 47830–47836. [CrossRef]

34. Fu, C.; Yao, M.; Chen, X.; Maulana, L.Z.; Li, X.; Yang, J.; Imasato, K.; Zhu, F.; Li, G.; Auffermann, G.; et al. Revealing the intrinsic electronic structure of 3D half-Heusler thermoelectric materials by angle-resolved photoemission spectroscopy. *Adv. Sci.* **2020**, *7*, 1902409. [CrossRef]
35. Buffon, M.L.C.; Laurita, G.; Verma, N.; Lamontagne, L.; Ghadbeigi, L.; Lloyd, D.L.; Sparks, T.D.; Pollock, T.M.; Seshadri, R. Enhancement of thermoelectric properties in the Nb–Co–Sn half-Heusler/Heusler system through spontaneous inclusion of a coherent second phase. *J. Appl. Phys.* **2016**, *120*, 75104. [CrossRef]
36. Dat, D.T.D.; Mahanti, S.D.; Pulikkoti, J.J. Electronic structure of Zr–Ni–Sn systems: Role of clustering and nanostructures in half-Heusler and Heusler limits. *J. Phys. Condens. Matter* **2014**, *26*, 275501.
37. Miyamoto, K.; Kimura, A.; Sakamoto, K.; Ye, M.; Cui, Y.; Shimada, K.; Namatame, H.; Taniguchi, M.; Fujimori, S.; Saitoh, Y.; et al. In-gap electronic states responsible for the excellent thermoelectric properties of Ni-based half-Heusler alloys. *Appl. Phys. Express* **2008**, *1*, 81901. [CrossRef]
38. Liu, Y.; Sahoo, P.; Makongo, J.P.A.; Zhou, X.; Kim, S.-J.; Chi, H.; Uher, C.; Pan, X.; Poudeu, P.F.P. Large enhancements of thermopower and carrier mobility in quantum dot engineered bulk semiconductors. *J. Am. Chem. Soc.* **2013**, *135*, 7486–7495. [CrossRef] [PubMed]
39. Guo, S.; Anand, S.; Brod, M.K.; Zhang, Y.; Snyder, G.J. Conduction band engineering of half-Heusler thermoelectrics using orbital chemistry. *J. Mater. Chem. A* **2022**, *10*, 3051–3057. [CrossRef]
40. Snyder, G.J.; Snyder, A.H.; Wood, M.; Gurunathan, R.; Snyder, B.H.; Niu, C. Weighted mobility. *Adv. Mater.* **2020**, *32*, e2001537. [CrossRef]
41. Ferluccio, D.A.; Smith, R.I.; Buckman, J.; Bos, J.-W.G. Impact of Nb vacancies and p-type doping of the NbCoSn-NbCoSb half-Heusler thermoelectrics. *Phys. Chem. Chem. Phys.* **2018**, *20*, 3979–3987. [CrossRef] [PubMed]
42. Kim, H.-S.; Gibbs, Z.M.; Tang, Y.; Wang, H.; Snyder, G.J. Characterization of Lorenz number with Seebeck coefficient measurement. *APL Mater.* **2015**, *3*, 041506. [CrossRef]

materials

MDPI

Article

Preparation and Properties of Boron Modified Phenolic Resin for Automotive Friction Materials

Jianrong Liu [1,*], Jialin Guo [1], Juanli Deng [1,*], Shangwu Fan [2], Xide Cai [2], Sijie Kou [2] and Shaobo Yang [2]

[1] School of Materials Science and Engineering, Chang'an University, Xi'an 710064, China; 2024131008@chd.edu.cn

[2] Science and Technology on Thermostructural Composite Materials Laboratory, Northwestern Polytechnical University, Xi'an 710072, China; shangwu_fan@nwpu.edu.cn (S.F.); cxd1513701031@163.com (X.C.); kousijie@mail.nwpu.edu.cn (S.K.); shaobo_yang@mail.nwpu.edu.cn (S.Y.)

* Correspondence: 2022131001@chd.edu.cn (J.L.); dengjl@chd.edu.cn (J.D.)

Abstract: To address the thermal fade problem of brake pads, a boron-modified phenolic resin with better temperature resistance is intended to be developed. By introducing B-O bonds and high-temperature-resistant units, the thermal decomposition temperature of the phenolic resin will be increased. The modified resin is obtained through a step-growth polymerization reaction and then incorporated into the brake pad formulation to be hot-pressed into samples. The thermal decomposition temperature of the resin is measured by TGA, and the thermal fade performance of the brake pad samples is analyzed through friction and wear experiments. The results show that the introduction of B-O bonds and the doping of nano-alumina have increased the thermal decomposition temperature of the phenolic resin to 480 °C, meeting the expectation. Brake pads molded with this resin as an adhesive showed significantly better thermal degradation than those made with ordinary phenolic resin. Meanwhile, during the braking process, the brake pads made from this resin form a complete and continuous friction film, demonstrating good mechanical properties and thermal fade performance. The wear amount under the entire braking test is also acceptable. In addition, an exploration of the thermal fade mechanism is carried out.

Keywords: automotive friction material; boron-modified phenolic resin; high heat resistance; friction and wear behavior

1. Introduction

The automotive manufacturing industry constitutes an important part of the global manufacturing sector, with the demand for automobiles increasing annually. The braking system is one of the most vital components of a vehicle. Brakes primarily convert kinetic energy into heat energy and other forms of energy through the frictional action of the rubbing pair [1–4].

C/C-SiC composites have low density, corrosion resistance, and excellent thermal stability and mechanical properties. When used as brake discs, they exhibit outstanding friction and wear performance [5–7]. Since 2001, C/C-SiC brake discs have been gradually applied to high-performance sports cars of manufacturers such as Porsche and Ferrari. At present, the semi-metallic brake pads paired with them hold a large market share. Their main components include reinforcing fibers, fillers, and organic binders, with phenolic resin materials being the main type of organic binders [8,9].

Currently, resin-based friction materials are widely utilized due to their characteristics such as low specific gravity, low braking noise, minimal damage to the mating disc, ease of molding and processing, and low cost. The phenolic resin features high hardness, high compressive strength, strong corrosion resistance, good wettability to fillers, and good heat resistance, as well as low cost and simple production process, and has thus always been the preferred choice for the resin matrix in friction materials [10–13]. However, with the increase in the operating speed and load of modern mechanical equipment and vehicles, the friction materials prepared from ordinary phenolic resin can no longer meet the usage requirements. The molecular chain of ordinary phenolic resin contains a large number of phenolic hydroxyl groups and methylene groups, which are prone to thermal decomposition. During high-temperature or emergency braking, the friction materials are liable to experience thermal fade, subsequently leading to a decrease in the braking force of the automobile and an increase in the wear amount, which significantly limits their application potential [14–17].

Consequently, enhancing the thermal decomposition temperature of the matrix is of great significance for improving the performance of resin-based friction materials. To address this issue, this paper will mainly focus on boron-modified phenolic resin and select three modification methods: (1) Introduction of B-O bonds via boric acid and doping with alumina nanoparticles; (2) Introduction of B-O bond and rigid benzene ring via phenylboronic acid and doping with nano-alumina at the same time; (3) Introduction of B-0 bond via boric acid and introduction of furan ring via furfural. Among them, Benzene ring can withstand higher temperatures. Furfural possesses two functional groups, namely the furan ring (conjugated double bonds: C=C-C=C) and the aldehyde group (C=O), and exhibits high reactivity and good heat resistance [18]. γ-nano-alumina has a uniform particle size distribution, high purity, good dispersion, a large specific surface area, and is inert to high temperatures

Semi-metal resin-based friction material samples are prepared by hot-pressing, and the friction and wear performance of the samples are tested using a scaled-down inertia test bench for automotive brakes. The aim is to compare and obtain the optimal phenolic resin modification scheme. Additionally, the surface morphology of the brake pads is observed using a scanning electron microscope, and the thermal fade theory is further analyzed in combination with chemical reactions.

2. Experiments

2.1 *Materials*

Phenol, 37% formaldehyde solution, furfural, sodium hydroxide (NaOH, granular), hexamethylenetetetramine, all analytically pure, supplied by Tianjin Damao Chemical Reagent Factory (Tianjin, China). Boric acid, phenylboronic acid, and γ-nano-alumina (γ-nano-Al_2O_3) were analytically pure and purchased from Shanghai Yien Chemical Technology Co., Ltd., Shanghai, China. All the above chemical reagents were used as received.

The friction materials were composed of steel fibers (average length 1.6 mm, average diameter 0.2 mm, Dezhou Fangyuan Steel Wool Fiber Co., Ltd., Dezhou, China) as the reinforcement and other components, as shown in Table 1.

Table 1. Friction material main formulations.

Components	Average Size (μm)	Mohs Hardness	Source
α-alumina	19	9	Henan Fengkai Refractories Co., Ltd. (Zhenzhou, China)
silicon dioxide	74	7	Heyuan Wanchuan Quartz Development Co., Ltd. (Heyuan, China)
graphite	250	2	Jiangxi Shuobang Material Technology Co., Ltd. (Fuzhou, China)
magnesia	44	5.5	Shanghai Ruhua Technology Co., Ltd. (Shanghai, China)
others	—	—	—

2.2. *Sample Preparation*

2.2.1. Preparation of Nano-Al_2O_3/BPR

In the mole ratio, phenol-formaldehyde-boronic acid = 1:1.2:0.2. Molten phenol, formaldehyde solution, and NaOH were added into a three-necked flask equipped with a spherical condenser tube and a mechanical stirring device [19]. The solution was stirred and heated to 65 °C~70 °C and held at that temperature for 1.5 h. In the second step, boric acid was added to this solution. The solution was heated to 105 °C~110 °C for 1 h. Thereafter, γ-nano-Al_2O_3 was added into the solution for a reaction of 30 min. The solution was cooled to room temperature. And the yellow-brown doped nano-alumina boron-modified phenolic resin mixed solution (nano-Al_2O_3/BPR) was obtained. Then, the mixed solution was put into a vacuum oven at 65 °C to dry for 24 h to remove excess water. Finally, the resin in the solid state is obtained. The samples doped with 1, 2, 3, 4, and 5 g of γ-nano-Al_2O_3 were named BPR-1, BPR-2, BPR-3, BPR-4 and BPR-5.

2.2.2. Preparation of Nano-Al_2O_3/PBPR

It was stated that phenylboronic acid was required to be ten percent of the total mass of the phenolic resin. Phenol, formaldehyde solution, and NaOH were added into a three-necked flask equipped with a stirrer and a condenser. The solution was stirred and heated to 65 °C~70 °C and held at that temperature for 1.5 h. Beyond, phenylboronic acid is added to this solution. And the solution was heated to 130 °C~140 °C for 1 h [20]. Then, γ-nano-Al_2O_3 was added into the solution for a reaction of 30 min. The solution was cooled to room temperature. A dark reddish-brown mixed solution of boron-modified phenolic resin doped with nano-alumina (nano-Al_2O_3/PBPR) is then obtained. Finally, this solution was removed in a vacuum to remove the water, and the solid resin was obtained. The samples doped with 1, 2, 3, 4, and 5 g of γ-nano-Al_2O_3 were named PBPR-1, PBPR-2, PBPR-3, PBPR-4, and PBPR-5.

2.2.3. Preparation of FPBPR

The formaldehyde-furfural solution ($n_{formaldehyde}$:$n_{furfural}$ = 1:0.1, 1:0.15, 1:0.2, 1:0.25, 1:0.3) was preformulated and loaded into a constant-pressure dropping funnel. Then, phenol and NaOH were added into a three-necked flask equipped with a stirrer, a condenser, and a funnel. The solution was warmed up to 70 °C~80 °C for 2.5 h of reaction. Thereafter, phenylboronic acid was added to this solution. The solution was stirred at 140 °C for 1.5 h. A black-brown solution of phenylboronic acid-modified phenolic resin with furfural as a cross-linking agent (FPBPR) was obtained at room temperature. Water in the resin solution was removed in a vacuum. Eventually, solid FPBPR was obtained. The samples doped with different masses of furfural were named FPBPR-1, FPBPR-2, FPBPR-3, FPBPR-4, and FPBPR-5.

2.2.4. Preparation of Friction Materials

The solid resin sample was ground to a fine powder by a mortar and pestle to be used as an adhesive and was mixed with other components in a small Y-mixer (100 r/min). The homogeneously mixed powder was poured into a mold and hot pressed in a flatbed vulcanizing machine. During hot pressing, the final temperature of the vulcanizing machine was set to 180 °C. The vulcanizing pressure was 43 MPa and the holding time was 20 min. Besides, three to five exhausts were carried out during the warming period. The square material obtained by hot pressing was cut in half into a pair of friction discs. The length was 40 mm. The width was 20 mm. The height was 10 mm. The pair of friction plates were heat-treated in an oven at 210 °C and held at that temperature for 24 h. Thus, the heat-treated samples were used for the following characterization. Samples of brake pads thermoformed with phenolic resin, BPR-2, PBPR-3, and FPBPR-4 as bonding agents were named Pad pr, Pad A, Pad B, and Pad C, respectively.

2.3. Testing and Characterization

2.3.1. Testing Methods

The bulk density and open porosity of the friction materials were measured by the Archimedes method. An analytical balance (AG135, Mettler Toledo, Zurich, Switzerland) with an accuracy of 0.0001 g was used to weigh the samples. The bulk density and open porosity were calculated by using Equations (1) and (2) as follows:

$$\rho_a = \frac{m_1 \rho_{H_2O}}{m_3 - m_2} \times 100\% \quad (1)$$

$$P_b = \frac{m_3 - m_1}{m_3 - m_2} \times 100\% \quad (2)$$

where ρ_a, P_b, m_1, m_2, and m_3, measured bulk density (g/cm^3), open porosity (100%), dry weight (g), weightlessness (g), and wet weight (g), respectively.

Brinell hardness was chosen in this study to characterize the macro-hardness of the friction material. The use of Rockwell hardness testers (HR-150A, Mage, Laizhou, China) was based on test method ISO 6506-1:2014. The size of the sample was 40 mm × 20 mm × 10 mm. The indenter diameter was selected as 10 mm. The test force was set as 9804 N for fiction materials and the duration time of the test force was set as 30 s. Thus, Brinell hardness was calculated by using Equation (3) as follows:

$$HBW = 0.102 \times \frac{2F}{\pi D\left(D - \sqrt{D^2 - d^2}\right)} \quad (3)$$

where *HBW*, *F*, *D*, and *d*, measured Brinell hardness, payload pressure (N), indenter diameter (mm), and indentation diameter (mm), respectively.

An electronic universal testing machine (ETM series, Wance, Shenzhen, China) was selected to measure the layer shear strength of friction materials. Based on the test standard GB/T 26739-2011, the size of the sample is 20 mm × 20 mm × 10 mm and the loading rate is 5 mm/min. The shear strength was calculated by using Equation (4) as follows:

$$\tau = \frac{F}{A} \quad (4)$$

where τ, *F*, and *A*, measured shear strength (MPa), maximum shear force (N), and sheared area (mm^2), respectively.

To simulate the real braking scenarios of automobiles, a scaled-down inertia test rig for automotive brakes (JF-122, Jilin Electrical and Mechanical Equipment Co., Ltd., Changchun,

China) was used to conduct braking test experiments. As shown in Figure 1, Figure 1a is the whole test bench setup, Figure 1b is the brake component part of it, and Figure 1c is a carbon-ceramic disc with its surface coated with Si-SiC.

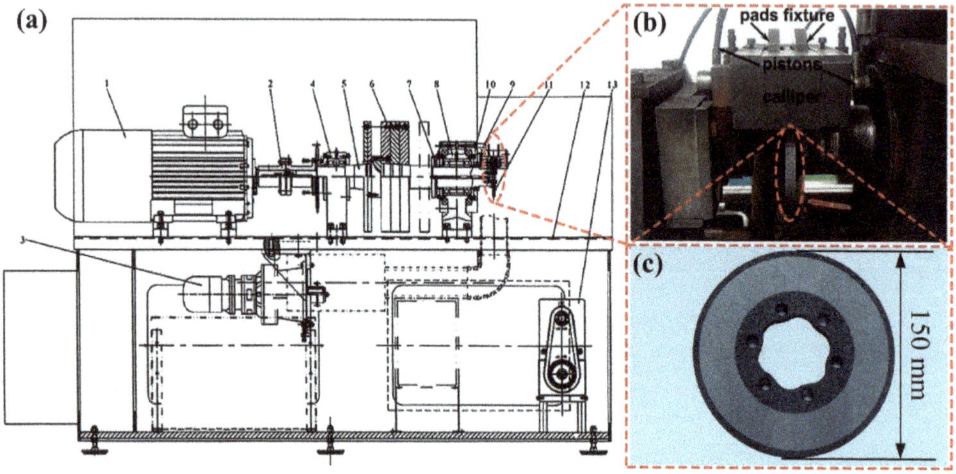

Figure 1. Schematic diagram of the test bench and physical drawing of brake disc. (**a**) schematic diagram [21]; (**b**) actual brake section [21]; (**c**) brake disc.

The brake pad size of the braking part is 40 mm × 20 mm × 10 mm, and the size of the brake disc is 150 mm in diameter and 10 mm in thickness. During the braking test, the surface temperature of the brake disc is measured using a contact-type K thermocouple. Next, the braking test program was determined by SAE-J2522 and detailed test conditions are summarized in Table 2. The frictional coefficient (COF) was calculated by using Equation (5) as follows:

$$\mu = \frac{M}{2 \times Arz \times R_m \times \eta \times P} \tag{5}$$

where μ, M, Arz, Rm, η, and P, measured COF, breaking torque (N·m), brake cylinder piston area (m^2), brake lever arm (m), braking efficiency (100%), braking strength (kPa). At the end of the braking test, the thickness of the brake discs and pads was measured using a micrometer with an accuracy of 0.001 mm. The brake disc was measured at 12 points and then averaged. Brake pads were measured at 5 points and then averaged.

Table 2. Test conditions of SAE-J2522 [21].

Sections	Test Name	Conditions
1	Green μ characteristic	80→30 km/h; 3.0 MPa; 30 stops
2	Burnish	80→30 km/h; 1.5–5.1 MPa; 64 stops
3	Characteristic value 1	80→30 km/h; 3.0 MPa; 6 stops
4	Speed/pressure sensitivity	40→5 km/h; 1.0–5.0 MPa; 5 stops
5	—	80→40 km/h; 1.0–5.0 MPa; 5 stops
6	—	120→80 km/h; 1.0–5.0 MPa; 5 stops
7	—	160→130 km/h; 1.0–5.0 MPa; 5 stops
8	—	180→150 km/h; 1.0–5.0 MPa; 5 stops
9	Characteristic value 2	80→30 km/h; 3.0 MPa; 6 stops
10	Cold braking	40→5 km/h; 3.0 MPa; 1 stop
11	Motorway braking	100→5 km/h; 40% deceleration; 1 stop
		160→10 km/h; 40% deceleration; 1 stop

Table 2. *Cont.*

Sections	Test Name	Conditions
12	Characteristic value 3	80→30 km/h; 3.0 MPa; 18 stops
13	1st fading	100→5 km/h; 40% deceleration; 15 stops
14	Recovery 1	80→30 km/h; 3.0 MPa; 18 stops
15	Temperature/pressure sensitivity 100 °C	80→30 km/h; 1.0–5.0 MPa; 5 stops
16	Temperature/pressure sensitivity 500 °C	80→30 km/h; 1.0–5.0 MPa; 9 stops;
17		80→30 km/h; 1.0–5.0 MPa; 5 stops
18	Recovery 2	80→30 km/h; 3.0 MPa; 18 stops
19	2nd fading	100→5 km/h; 40% deceleration; 15 stops
20	Recovery 2	80→30 km/h; 3.0 MPa; 18 stops

2.3.2. Characterization

The structural composition of the modified resin was characterized by fourier transform infrared spectroscopy (FT-IR, TENSOR II, Bruker, Wurzbach, Germany). Prior to analysis, all samples were obtained by the compression method and transmission mode is used for testing. In this study, the thermal weight loss of the modified resin was tested by a thermogravimetric analyzer (TGA, SDT650, TA, USA). The test program was set to increase the temperature of the sample from room temperature to 800 °C in an air atmosphere at a rate of 10 °C/min. DSC isothermal curves of modified resins were obtained by differential scanning calorimetry (DSC, DSC300C, DZ, Beijing, China). The samples were rapidly heated up to 150 °C/180 °C/210 °C in an air atmosphere at a rate of 50 °C/min and then held at that temperature for half an hour. After braking tests, the surface morphology of brake discs and pads was characterized by an optical microscope (OM, YM710TR, Prague, Czech Republic) and tungsten filament scanning electron microscope (SEM, Vega4, Tescan, Brno, Czech Republic), respectively.

3. Results and Discussion

3.1. Properties and Characterization of Resins

3.1.1. Structure of Modified Phenolic Resins

Due to the relationship between the infrared characteristic peaks of nano-sized aluminum oxide and its specific surface area and dehydration state [22,23], the characteristic peaks are complex. Since γ-nano-Al_2O_3 was only physically blended into the phenolic resin in this experiment, the infrared characteristic peaks of nano-alumina are not analyzed.

Figure 2a shows the FTIR spectra of PR, BPR, and cured BPR. Figure 2b shows the FTIR spectra of PR, PBPR, and cured PBPR. The main absorption peaks are assigned as follows: 3350 cm^{-1} (C-OH), 2924 cm^{-1}, 1085 cm^{-1}, 755 cm^{-1} (C-H), 1600 cm^{-1}, 1486 cm^{-1} (C=C, benzene ring), 1436 cm^{-1} (B-O, coordinate bond), 1365 cm^{-1} (B-O), and 1230 cm^{-1} (C-O). Notably, the peak intensities at 1085 cm^{-1} and 755 cm^{-1} (C-H stretching vibrations of the benzene ring) and at 1600 cm^{-1} and 1486 cm^{-1} (C-C stretching vibrations of the benzene ring) remain unchanged.

In comparison with PR, new absorption peaks appear at 1436 cm^{-1} and 1365 cm^{-1} in the spectra of BPR and PBPR, indicating the presence of B-O in the molecular structure. Additionally, the peak intensities at 3350 cm^{-1} (O-H stretching of phenolic hydroxyl) and 1230 cm^{-1} (C-O stretching of alcoholic hydroxyl) decrease. This reduction is due to the esterification reaction between B-OH in boric acid and some phenolic and alcoholic hydroxyl groups, forming borate esters or phenylboronic esters, which leads to a reduction in the content of phenolic and alcoholic hydroxyl groups.

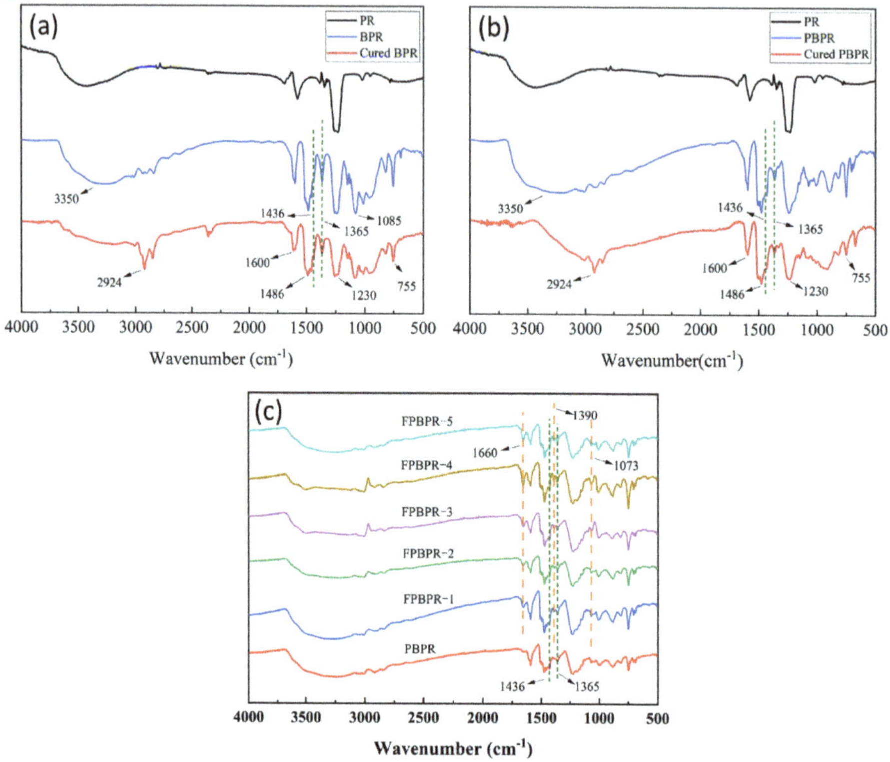

Figure 2. FTIR spectra of modified phenolic resins (**a**) BPR; (**b**) PBPR; (**c**) FPBPR.

Compared to BPR and PBPR, the peak intensities at 2924 cm^{-1} (C-H stretching vibrations of alkanes) in cured BPR and cured PBPR increase, indicating that during the curing process, some B-OH and phenolic hydroxyl groups continue to undergo dehydration condensation with alcoholic hydroxyl groups, leading to an increase in the number of methylene groups within the molecular chains. Therefore, it can be known that, regardless of whether the reaction is with boric acid or phenylboronic acid, B-O is ultimately introduced into the phenolic resin through esterification reactions, and the reaction becomes more complete after curing.

Figure 2c displays the FTIR spectra of PBPR and FPBPR with different furfural contents. Both PBPR and FPBPR show absorption peaks at 1436 cm^{-1} and 1365 cm^{-1}, confirming the incorporation of B-O. Compared to PBPR, FPBPR exhibits absorption peaks at 1660 cm^{-1} (C=O stretching vibration of furfural) and 1390 cm^{-1} (C=C stretching vibration of furfural). The intensity of these peaks increases with higher furfural content, which can be attributed to the ortho or para addition reaction between the -CHO group on the furan ring and phenolate ions, incorporating the furan ring into the molecular chain. Furthermore, the intensity of the peak at 1073 cm^{-1} (C-O-C stretching vibration) also increases with higher furfural content. This results from the intermolecular dehydration between the alcoholic hydroxyl group on the furan ring and the phenolic hydroxyl group, forming more C-O-C bonds. This further demonstrates the incorporation of furfural into the molecular chain.

3.1.2. Thermogravimetric Properties of Modified Phenolic Resins

The thermogravimetric analysis (TGA) of modified phenolic resins was conducted in an air atmosphere, and the thermogravimetric curves for the three types of modified phenolic resins are shown in Figure 3. Figure 3a presents the TGA curves for boric acid-modified phenolic resins doped with different contents of γ-nano-Al_2O_3. Generally, a noticeable decrease in weight occurs only after the temperature exceeds 450 °C, with the initial weight loss rate remaining below 15%. The carbon residue at 800 °C ranges from 25% to 45%, with some of this attributed to the presence of nano alumina. This behavior is due to the introduction of B-O bonds in the molecular chain of the phenolic resin, which possesses higher bond dissociation energies compared to C-C, C-O, and C-H bonds. The higher energy requirement for breaking B-O bonds makes them less prone to decomposition in high-temperature environments. Additionally, the incorporated γ-nano-Al_2O_3 exhibits high thermal stability. The synergistic effect of these factors leads to an increased thermal decomposition temperature for the modified phenolic resins compared to conventional phenolic resins (280 °C~350 °C).

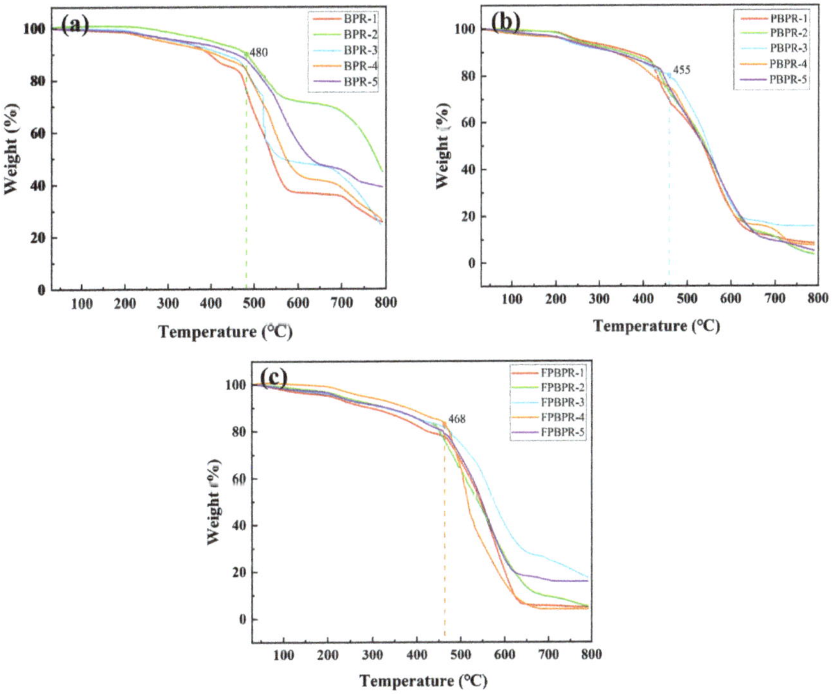

Figure 3. TG curves of weight loss for the modified phenolic resins. (**a**) BPR; (**b**) PBPR; (**c**) FPBPR

Examining the individual samples, the thermal decomposition temperatures of BPR-2, BPR-3, BPR-4, and BPR-5 are all higher than that of BPR-1. Furthermore, the thermal decomposition temperatures of BPR-1, BPR-3, BPR-4, and BPR-5 exhibit a gradually increasing trend, indicating that the incorporation of γ-nano-Al_2O_3 enhances the thermal stability of the resin. The amount of γ-nano-Al_2O_3 doping is roughly proportional to the thermal decomposition temperature of the resin. Among these, BPR-2 demonstrates the best thermal stability, initiating thermal decomposition only at temperatures exceeding 480 °C. Its weight loss rate is 10%, and the residual carbon rate at 800 °C is 45%. This optimal performance may result from the ideal ratio of γ-nano-Al_2O_3, where steric hindrance effects are minimized, allowing for a uniform distribution within the resin's molecular chains.

Figure 3b shows the TGA curves for phenylboronic acid-modified phenolic resins doped with varying amounts of γ-nano-Al_2O_3. A significant weight loss trend becomes evident after the temperature exceeds 415 °C, with weight loss rates remaining below 25%. The carbon residue at 800 °C ranges between 5% and 20%. The thermal decomposition temperatures of PBPR-2, PBPR-3, PBPR-4, and PBPR-5 are all higher than that of PBPR-1. Among these, the thermal decomposition temperatures of PBPR-1, PBPR-2, and PBPR-5 show a gradual increase, while the weight loss rate during decomposition exhibits an increasing trend. This suggests that both the introduction of B-O bonds and the incorporation of γ-nano-Al_2O_3 contribute to enhancing the thermal decomposition temperature. However, the method of introducing B-O bonds via phenylboronic acid, which contains a benzene ring, leads to significant steric hindrance. As the amount of γ-nano-Al_2O_3 increases, the effective participation of phenylboronic acid in the reaction diminishes, resulting in a reduced number of high-temperature-resistant B-O bonds and potentially leading to decreased thermal stability of the resin. Notably, PBPR-3 has the highest thermal decomposition temperature of 455 °C. Its weight loss rate is 20%, and the residual carbon rate at 800 °C is 20%. This may indicate a suitable balance between the introduction of B-O bonds and the doping of γ-nano-Al_2O_3.

Figure 3c illustrates the TGA curves for phenylboronic acid-modified phenolic resins crosslinked with different amounts of furfural. The thermal decomposition temperatures of these samples are relatively similar, ranging between 440 °C and 460 °C, except for FPBPR-2, which shows no significant differences compared to the others. As the furfural content increases, there are no discernible trends in either the weight loss rate or carbon residue. This lack of correlation may be attributed to the presence of a furan ring in furfural and a benzene ring in phenylboronic acid, both of which impose steric hindrance effects that limit their interaction. Consequently, the effective participation of both the benzene ring and B-O bonds, as well as the furan ring, is restricted, thus limiting the further enhancement of the resin's thermal stability. Among these samples, FPBPR-4 exhibits the highest thermal decomposition temperature at 468 °C.

3.1.3. Curing of Modified Phenolic Resins

Figure 4 presents the DSC isothermal curves of three types of modified phenolic resins at different temperatures, along with the corresponding degree of cure curves derived from the fittings. The samples BPR-2, PBPR-3, and FPBPR-4 were subjected to rapid heating at 150 °C, 180 °C, and 210 °C, respectively, maintaining each temperature for 30 min to obtain the DSC isothermal curves, as shown in Figure 4a–c,e–g,i–k. All curves exhibited a distinct exothermic peak within the first 10 min, attributed to the curing reaction of the resin upon reaching a specific temperature, releasing heat. Following this, the curing reaction concluded within 5 to 10 min, resulting in a plateau in the curves.

By performing baseline fitting, integrating the areas, and calculating the area ratios, the degree of cure curves for BPR-2, PBPR-3, and FPBPR-4 at various temperatures were ultimately fitted, as depicted in Figure 4d,h,l. Overall, the curing process was completed within 10 min for all samples. Although the curing rate slightly decreased with increasing temperature, the time required for complete curing correspondingly reduced. This information can be utilized to design the curing process for the resin. However, due to the small sample size used in the experimental tests, which may slightly differ from actual conditions, appropriate adjustments to the curing time of the resin should be taken.

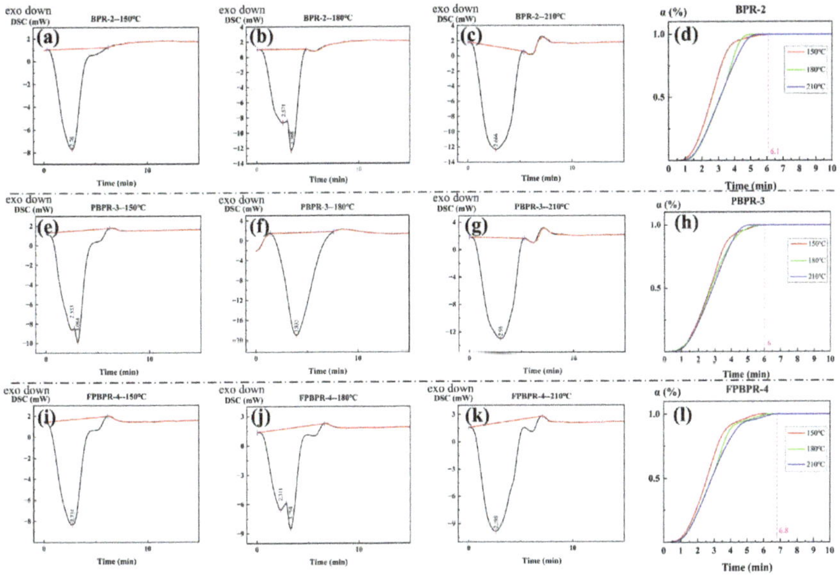

Figure 4. DSC isothermal curves and corresponding curability curves of modified phenolic resins. (**a**,**e**,**i**) 150 °C isothermal curve of BPR-2, PBPR-3, FPBPR-4; (**b**,**f**,**j**) 180 °C isothermal curve of BPR-2, PBPR-3, FPBPR-4; (**c**,**g**,**k**) 210 °C isothermal curve of BPR-2, PBPR-3, FPBPR-4; (**d**,**h**,**l**) Curing degree curve of BPR-2, PBPR-3, FPBPR-4.

3.2. Characterization of Brake Pads

3.2.1. Brinell Hardness and Shear Strength

Shear strength and the Brinell hardness of the brake pads made from phenolic resin and three types of boron-modified phenolic resins are shown in Figure 5a,b. Shear strength and Brinell hardness of the Pad pr are the largest, indicating good bonding properties. However, too strong bonding properties lead to an increase in the denseness of the sheet, which results in an increase in the coefficient of friction. Therefore, the other three pads may be the most suitable in meeting the requirements of brake pads. The Brinell hardness of Pad A is 46.87, and the unit shear stress is 8.25 MPa. Whether it is the Brinell hardness or the shear strength, the values of Pad A are higher than those of Pad B and Pad C. And they meet the national standards, which shows that Pad A is better than Pad B and Pad C in mechanical properties. This has a positive effect on Pad A in the friction and wear tests. This is mainly because the resin used in Pad A has better bonding strength.

It can be seen from Table 3 that the Pad pr has the highest density. This also corresponds to the fact that the shear strength and Brinell hardness of the Pad pr in picture 5 are the largest. As can be seen in Table 3, the higher the open porosity, the lower the density of the sheet, which means that there are more pores on the surface and inside the pad. When the sheet is subjected to shear force, the stress cannot be well dispersed, resulting in localized stress concentration, which makes the pad more susceptible to damage and the shear strength decreases. After excluding its effect on the frictional coefficient, the density of Pad A is greater than that of Pad B and Pad C. Meanwhile, the open porosity of Pad A is smaller than that of Pad and Pad C.

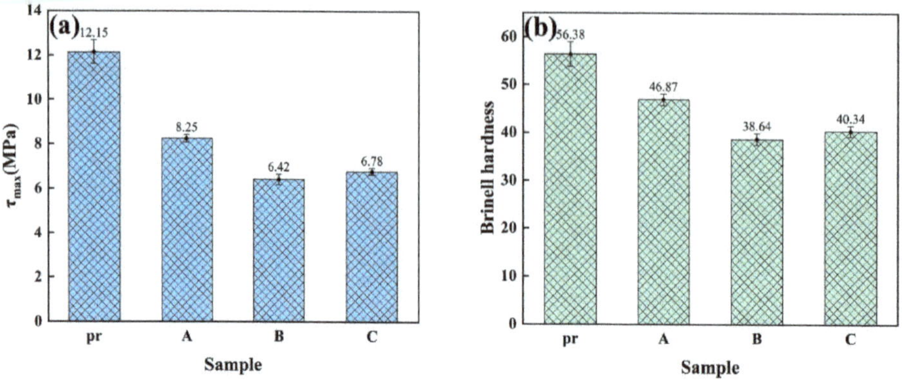

Figure 5. (**a**) shear strength of brake pads; (**b**) Brinell hardness of brake pads.

Table 3. Density and open porosity of pads.

Simples	Density (g/cm^3)	Open Porosity (%)
pr	3.83	9.88
A	3.54	10.76
B	3.35	10.98
C	3.47	10.84

3.2.2. Friction and Wear Properties

The thermal fade curves of the four types of pads are presented in Figure 6. As can be seen from Figure 6a, the frictional coefficient of Pad pr continues to decrease with increasing temperature, and the thermal fading phenomenon is very obvious. From Figure 6b, it can be known that with the increase of temperature, the frictional coefficient of Pad A firstly decreases and then slowly recovers to above 0.4. From Figure 6c,d, the frictional coefficients of Pad B and Pad C gradually recover as the temperature increases after decreasing to a certain range.

Combined with the performance of the four types of brake pads in the thermal fade stage and the above test data, it indicates that boron-modified phenolic resins are successfully demonstrated on brake pads. In a word, boron-modified phenolic resins largely improve the deficiencies of ordinary phenolic resins in the thermal fading of brake pads.

Figure 7 illustrates the amount of wear loss for the three types of brake pads. From the Figure 7, it can be seen that Pad A has the smallest wear loss of 0.728 mm, while Pad B and Pad C have much larger wear loss than Pad A. The combination of thermal degradation and wear loss suggests that the selection of BPR-2 as the brake pad binder is the most suitable one.

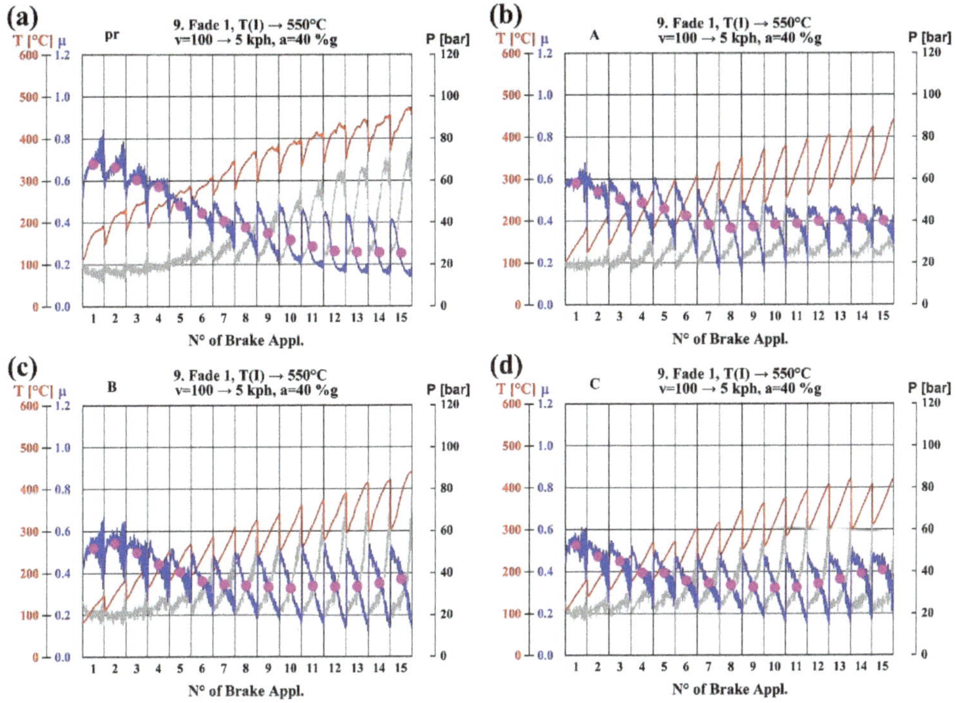

Figure 6. (**a**) The fade curve of Pad pr; (**b**) The fade curve of Pad A; (**c**) The fade curve of Pad B; (**d**) The fade curve of Pad C. (Grey line correspondent to P [bar]).

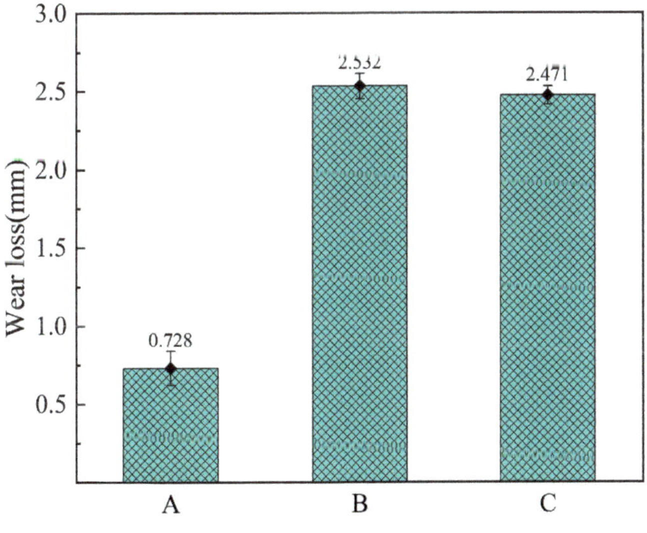

Figure 7. Histogram of brake pads wear loss.

3.3. Thermal Fade

To explore the causes of the thermal fade of the brake pads, the thermal fade braking process of the rubbing pair was divided into three stages:

(1) The first stage ends at the fifth braking, with a temperature of 350 °C, corresponding to the initial descending stage of the thermal fade curve.

(2) The second stage ends at the tenth braking, with a temperature of 420 °C, corresponding to the point where the thermal fade curve drops to its lowest level.

(3) The third stage ends at the fifteenth braking, with a temperature of 480 °C, corresponding to the recovery stage in the later period of the thermal fade curve.

3.3.1. Surface Morphology of Brake Discs

Figure 8 shows the surface morphologies of the brake disc at three stages. The surface of this disc is mainly composed of Si and SiC. In Figure 8a, the light gray area is composed of SiC, the bright white area is Si, and the golden yellow part is Fe_2O_3. Some areas on the surface of the brake disc are covered by a transfer layer formed by the wear debris of the friction material. The blue arrows indicate the scratches on the disc surface and the braking direction of the rubbing pair. In Figure 8b, a new blue area appears, and the composition of this part is Fe_3O_4, and most areas on the surface of the brake disc have been covered by the transfer layer. In Figure 8c, the original appearance of the disc itself can hardly be seen but is instead replaced by a complete and continuous friction film.

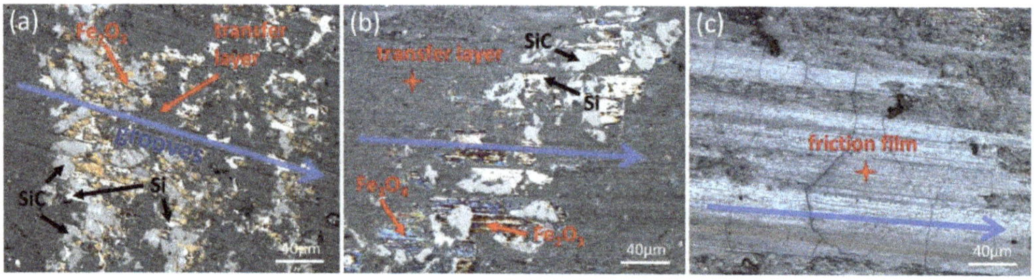

Figure 8. Surface morphology of brake disc at different stages of thermal fade. (**a**) Surface morphology of the first stage; (**b**) Surface morphology of the second stage; (**c**) Surface morphology of the third stage.

From the first stage to the second stage, the transfer layer gradually covers the disc surface. However, compared with the process from the second stage to the third stage, the covering speed of the transfer layer is not very fast. This is because the temperature in the early stage of thermal fade is lower than that in the later stage. In the early stage, the wear debris of the friction material will transfer to the surface of the brake disc, and at the same time, the wear debris of the brake disc will also transfer to the surface of the friction material (adhesion mechanism). In the later stage of thermal fade, as the temperature rises, the transfer is mainly that of the wear debris of the friction material to the surface of the brake disc (diffusion mechanism) [24]. It can be inferred from this that the friction film was not fully formed in the early and middle stages, which led to the occurrence of thermal fade. In the later stage, a complete and continuous friction film was formed, improving the friction coefficient and making it stable.

3.3.2. Surface Morphology of Brake Pads

The surface morphologies of the brake pads in the first stage of thermal fade are presented in Figure 9. In Figure 9a, according to face scanning results in picture (b), most areas form the main contact platforms composed of steel fibers, while a small number of areas accumulate the secondary contact platforms by the accumulation areas of graphite with larger particle sizes and wear debris. However, each contact platform is relatively independent and has a certain height difference. In Figure 9c, the larger particle phases that can be observed are known to be SiO_2 and MgO respectively through EDS point scanning analysis of Figure 9e, because they have larger Mohs hardness and particle sizes. Analyzing Region C, the wear debris is mainly composed of graphite, steel fibers, oxides, and other formulation materials such as MgO and Al_2O_3. Analyzing Region D, its composition is similar to that of Region C, indicating that some steel fibers have friction materials adhered to their surfaces. When the red box area in Figure 9c is magnified, it can be seen that it is covered with a lot of debris. Through point scanning analysis, it is known that the debris is mainly composed of iron filings and a small amount of graphite, which also shows that during the process of the steel fibers extending outward, the shed iron filings and other debris particles continuously accumulate and form a secondary contact platform. From Figure 9f, it can be seen that there are slight scratches on the surface of the brake pad, accompanied by a yellow substance, which is presumed to be Fe_2O_3 [25,26], and it is possible that an oxide film formed on the surface of the brake pad.

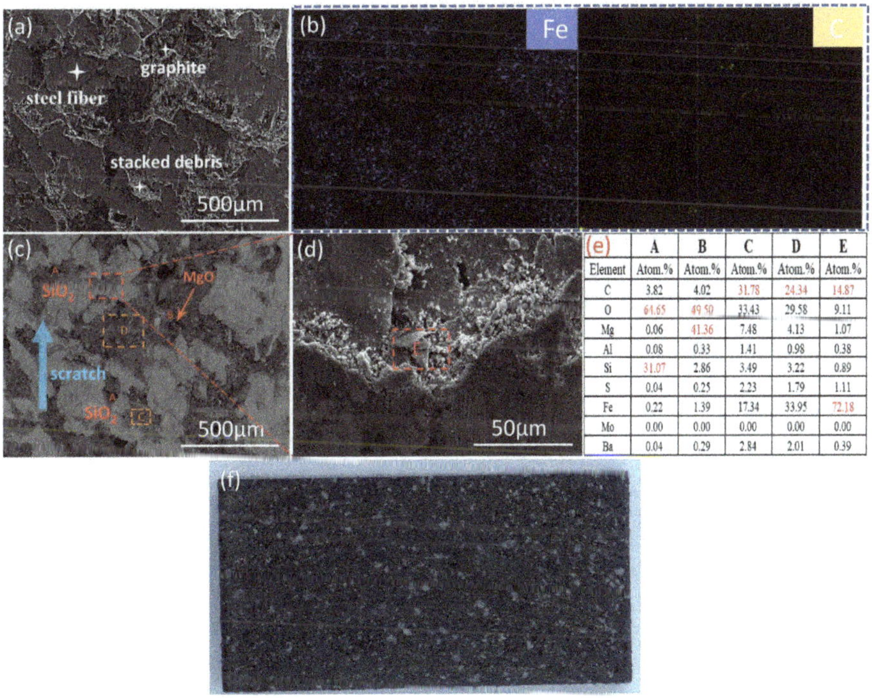

(e)	A	B	C	D	E
Element	Atom.%	Atom.%	Atom.%	Atom.%	Atom.%
C	3.82	4.02	31.78	24.34	14.87
O	64.65	49.50	33.43	29.58	9.11
Mg	0.06	41.36	7.48	4.13	1.07
Al	0.08	0.33	1.41	0.98	0.38
Si	31.07	2.86	3.49	3.22	0.89
S	0.04	0.25	2.23	1.79	1.11
Fe	0.22	1.39	17.34	33.95	72.18
Mo	0.00	0.00	0.00	0.00	0.00
Ba	0.04	0.29	2.84	2.01	0.39

Figure 9. Surface morphology of brake pads at the first stage. (**a**,**d**) Secondary electron scans at different magnifications; (**c**) Backscattered electron scan; (**f**) Macroscopic surfaces; (**b**,**e**) EDS results.

Figure 10 shows the surface morphologies of the brake pads in the second stage of thermal fade. As shown in Figure 10a, the connection between the main contact platform composed of steel fibers and the secondary contact platform composed of wear debris becomes tighter, with no obvious height difference, and the whole look much smoother. Due to the high Mohs hardness of SiO_2 and MgO, which are not prone to wear, the relatively obvious particle phases can still be seen in Figure 10b. The EDS point scanning results of Figure 10e indicate that the components of Region C and Region D are still mainly wear debris. Compared with the first stage of thermal fade, more areas on the steel fibers are adhered to by the friction materials and the amount of graphite increases. Figure 10c is the magnification of the red box area in Figure 10b. The wear debris accumulation area in the upper part is relatively flat, and the contact between the abrasive grains is close. In the wear debris accumulation area E, due to the insufficiently compact accumulation, cracks are formed after being subjected to stress. It can be presumed that these changes mentioned above have a positive impact on the further reduction of COF. Figure 10d shows that the scratches on the surface of the brake pad deepened. Upon closer inspection, it can also be found that a small amount of blue substance where the orange arrows point appeared on the surface of the brake pad, and it is presumed that a new oxide film started to form on its surface.

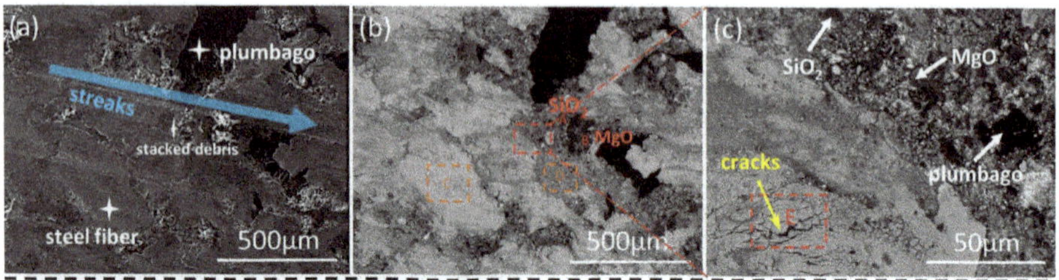

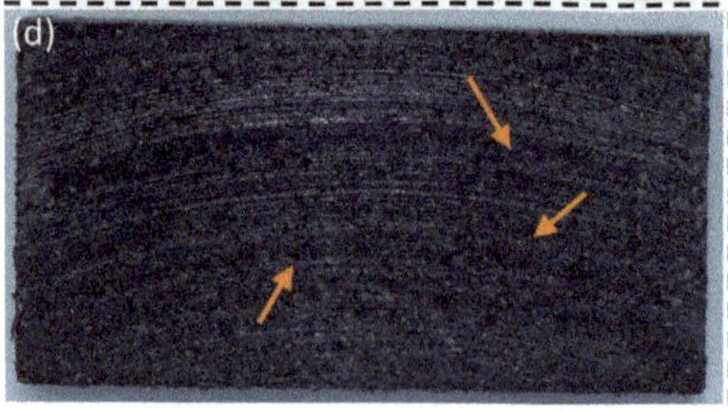

(e) Element	A Atom.%	B Atom.%	C Atom.%	D Atom.%	E Atom.%
C	6.79	3.89	35.74	27.62	51.70
O	60.86	48.78	17.74	24.12	13.84
Mg	0.96	45.08	0.64	4.22	0.97
Al	0.06	0.16	0.41	1.23	0.44
Si	29.55	0.56	0.63	1.84	0.52
S	0.02	0.20	0.57	0.60	0.58
Fe	1.72	1.04	44.00	39.50	31.39
Mo	0.00	0.00	0.00	0.11	0.06
Ba	0.04	0.29	0.27	0.76	0.50

Figure 10. Surface morphology of brake pads at the second stage. (**a**) Secondary electron scan; (**b**,**c**) Backscattered electron scans at different magnifications; (**d**) Macroscopic surfaces; (**e**) EDS results.

Figure 11 shows the surface morphologies of the brake pads in the third stage of thermal fade. As shown in Figure 11a, the connections between the various contact platforms are so tight that it is almost impossible to distinguish their boundaries, and a complete and continuous friction film is formed. Through point scanning analysis, the friction materials are evenly dispersed on the surface of the brake pad. In Figure 11c, through EDS point scanning analysis of Figure 11f, Region E is an area of wear debris accumulation. It can be seen that the connections between the wear debris are getting closer, forming a smooth surface. In addition, there are many white shiny particles distributed on the friction interface. Analyzing their composition, they are relatively hard particles containing a large amount of iron filings. From this, it can be presumed that in the third stage of thermal fade, three-body abrasive wear mainly occurs between the friction interfaces, and obvious scratches can be seen in Figure 11d. In Figure 11d,e, large areas of blue substances appear on the surface of the brake pads. Judging from the temper colors, they are mainly Fe_3O_4. Along the direction of the red arrow, the temperature gradually rises, indicating that the temperature of the friction pads is mainly concentrated in the central area.

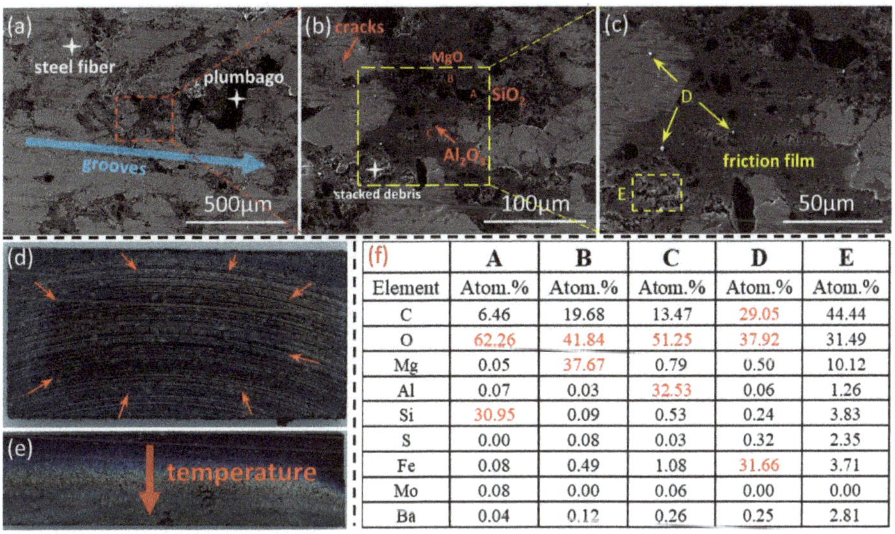

Element	A Atom.%	B Atom.%	C Atom.%	D Atom.%	E Atom.%
C	6.46	19.68	13.47	29.05	44.44
O	62.26	41.84	51.25	37.92	31.49
Mg	0.05	37.67	0.79	0.50	10.12
Al	0.07	0.03	32.53	0.06	1.26
Si	30.95	0.09	0.53	0.24	3.83
S	0.00	0.08	0.03	0.32	2.35
Fe	0.08	0.49	1.08	31.66	3.71
Mo	0.08	0.00	0.06	0.00	0.00
Ba	0.04	0.12	0.26	0.25	2.81

Figure 11. Surface morphology of brake pads at the third stage. (**a**) Secondary electron scan; (**b**,**c**) Backscattered electron scans at different magnifications; (**d**,**e**) Macroscopic surfaces; (**f**) EDS results.

From the first stage to the third stage of thermal fade, due to the good ductility of steel fibers, they gradually spread out on the friction surface as the temperature rises. Under the action of stress, the particle size of the wear debris decreases and tightly accumulates to form a secondary contact platform. Some of the wear debris will adhere to the surface of the steel fibers while wrapped with a large amount of graphite, improving the lubrication. Then, the primary contact platform and the secondary contact platform jointly form a complete and continuous friction film. With the assistance of the surface oxide film, the coefficient of friction will recover to some extent until it stabilizes.

3.3.3. Analysis of Thermal Fade

The diagram of the thermal fade model is presented in Figure 12. Gray and black parts represent brake disc, orange parts represent brake pad. Figure 12a shows that in the pre-stage of thermal fade, the micro-convex body on the surface of the brake disc and the surface of the brake pad are undergoing a "plow effect", which destroys the surface structure of the brake pad. Figure 12b shows the post stage of thermal fade, where a friction film is formed on the surface of both the brake pad and the brake disk. The specific mechanism is analyzed as follows. During the thermal fade stage, the temperature of the brake pads will increase with the increase in the number of braking operations.

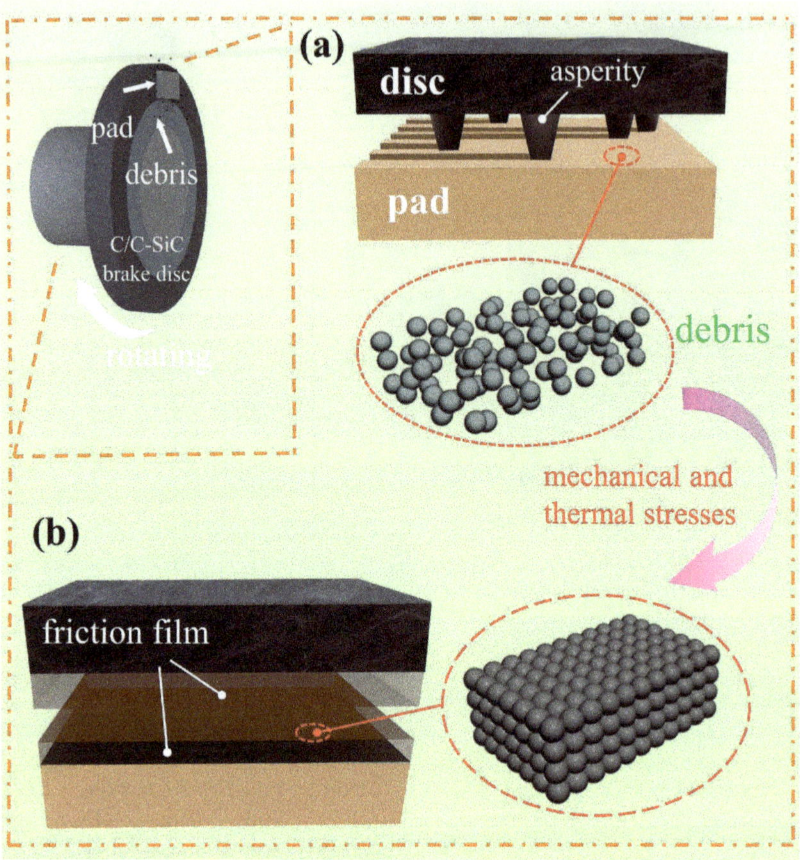

Figure 12. Modeling diagram for thermal fade.

In the preliminary stage, the displayed temperature does not reach the thermal decomposition temperature of the resin. In principle, the frictional coefficient should not drop significantly. However, an external thermocouple was used in this experiment, and the actual temperature at the contact surface between the brake disc and the brake pads is higher than the displayed temperature and may have already reached the thermal decomposition temperature of the resin. Consequently, the resin begins to decompose, the adhesion between components is lost, and the surface of the brake pads becomes loose and easily damaged by the hard asperities on the surface of the brake disc. Meanwhile, the frictional coefficient decreases accordingly.

In the post stage, the wear debris shed from the surface of the brake pads is transferred to the surface of the brake disc, resulting in adhesive wear. The frictional coefficient continues to decline. At the same time, as the temperature rises, the residual wear debris on the surface of the brake disc is transferred back to the surface of the brake pads [24]. Under the action of mechanical stress and thermal stress, the wear debris on the surface of the brake pads is gradually compacted, and the bonding relationship between the debris becomes close, forming a stable and flat friction film on the surface of the brake pads and disc, and the coefficient of friction is restored and stabilized.

4. Conclusions

To improve the thermal fade performance of the brake pads, this study started by increasing the thermal decomposition temperature of the phenolic resin. The means of introducing B-O bonds, rigid cyclic structures, and physically doping high-temperature-resistant components were adopted to modify the phenolic resin. The thermogravimetric results showed that the phenolic resin modified jointly by boric acid and γ-nano-Al_2O_3 had a higher thermal decomposition temperature, reaching about 480 °C. Moreover, the results of the friction and wear tests indicated that thermal fade performance of brake pads with boron-modified phenolic resin as binder is better than that of brake pads with ordinary phenolic resin as binder. This demonstrates that the improvement of the resin's temperature resistance indeed has a positive effect on the thermal fade performance of the brake pads. Meanwhile, it also provides a basis for how to maintain the stability of the friction pad performance under higher braking temperatures in subsequent studies.

Furthermore, in the analysis of the causes of thermal fade, it was found that the reason for the decrease in the friction coefficient in the early stage is most likely still related to the thermal decomposition temperature of the resin. However, due to the lack of measurement of the actual temperature at the contact surface between the brake disc and the brake pads, a definite conclusion cannot be reached.

Currently, the thermal decomposition temperature of the resin studied in this experiment is still lower than 500 °C, and the problem of the temperature resistance of the resin has not been fundamentally solved. Nevertheless, this experiment verified that the increase in the thermal decomposition temperature of the resin has a positive impact on the thermal fade performance. This provides more ideas and solutions for the subsequent improvement of the friction and wear performance of the friction lining.

Author Contributions: Conceptualization, J.L. and S.F.; methodology, X.C. and S.K.; validation, J.G. and J.D.; data curation, J.G.; writing—original draft preparation, J.L.; writing—review and editing, J.D.; supervision, S.Y.; funding acquisition, S.F. All authors have read and agreed to the published version of the manuscript.

Funding: This research was funded by the Natural Science Foundation of China grant number 52272090. And the APC was funded by Shangwu Fan. Besides, we would also like to thank the Analytical &Testing Center of Northwestern Polytechnical University and Shaanxi Materials Analysis and Research Center for their support of equipment.

Institutional Review Board Statement: Not applicable.

Informed Consent Statement: Not applicable.

Data Availability Statement: Data cannot be disclosed due to privacy.

Conflicts of Interest: The authors declare no conflict of interest.

References

1. Ling, J.H.; Luan, D.C.; Hu, Z.H.; Liu, L.; Ren, Y.; Zhou, X.; Wang, Z. Study on the effects of different kinds of phenolic resin binders on the properties of friction materials. *China Plast. Ind.* **2022**, *50*, 68–72.
2. Awe, S.A. Developing material requirements for automotive brake disc. *Mod. Concepts Mater. Sci.* **2019**, *2*, 000531.
3. Kosbe, P.; More, C. Performance of a NAO friction material for disc brake. *Natl. J. Adv. Build. Sci. Mech.* **2010**, *1*, 1–5.
4. Kumar, V.V.; Kumaran, S.S. Friction material composite: Types of brake friction material formulations and effects of various ingredients on brake performance-A review. *Mater. Res. Express* **2019**, *6*, 082005.
5. Dong, C.Y.; Deng, J.L.; Fan, S.W.; Kou, S.; Yang, S.; Huang, R.; Zhang, Y.; Mao, Y. Tribological performance and composition optimization of semi-metallic friction materials applied for carbon ceramic brake disc. *Ceram. Int.* **2024**, *50*, 19660–19670.
6. Ma, X.; Fan, S.W.; Luan, C.H.; Li, W.; Deng, J.; Cheng, L.; Zhang, L. Effect of Cu addition on the braking performance of Fe-Si alloy modified C/CSiC brake materials. *Wear* **2021**, *477*, 203851.

7. Fan, S.W.; Ning, Y.F.; Ma, X.; Yang, C.; He, L.; Deng, J.; Cheng, L.; Zhang, L. The effects of high-temperature annealing on microstructure and performance of Cf/C-SiC composites modified by $FeSi_2$. *Ceram. Int.* **2018**, *44*, 15200–15207.
8. Ma, X.; Luan, C.H.; Fan, S.W.; Deng, J.; Zhang, L.; Cheng, L. Comparison of braking behaviors between iron- and copper-based powder metallurgy brake pads that used for C/C-SiC disc. *Tribol. Int.* **2021**, *154*, 106686.
9. Jiang, H.; Wang, C.X. Foreign friction material products and research progress. *Adv. Mater. Ind.* **2021**, *3*, 2–6.
10. Gurunath, P.V.; Bjiwe, J. Friction and wear studies on brake-pad materials based on newly developed resin. *Wear* **2007**, *263*, 1212.
11. Cai, P.; Qi, S.Q.; Leng, Y. Research progress of resin based friction materials. *J. Huaiyin Inst. Technol.* **2018**, *27*, 9–14.
12. Wang, Z.F.; Zhao, L.; Shang, P. Effect of compound modified phenolic resin on the performance of brake friction materials. *China Pet. Chem. Stand. Qual.* **2019**, *39*, 121–122.
13. Cong, P.H.; Wu, X.Y.; Bu, J.; Li, T.-S. Progress in research of organic friction materials for automobile and train braking. *J. Tribol.* **2011**, *31*, 88–96.
14. Lai, X.H.; Wan, X.; Gan, G.L. Research status and development of automotive brake materials. *Mater. Res. Appl.* **2020**, *14*, 240–245.
15. Wang, X.; Liang, X.L. Research progress on high temperature resistant phenolic resins. *Chem. Adhes.* **2021**, *43*, 304–306.
16. Sun, G.X. Research progress on modification of heat resistance of phenolic resin. *New Technol. New Prod. China* **2016**, *22*, 79–80.
17. Tang, K.H.; Zhang, A.L.; Ge, T.J.; Liu, X.; Tang, X.; Li, Y. Research progress on modification of phenolic resin. *Mater. Today Commun.* **2021**, *26*, 101879.
18. Li, W.Q.; Yao, J.Q.; Fan, Y.X.; Yu, C. Preparation and properties of furfural-phenol-formaldehyde co-condensation resin. *Plast. Sci. Technol.* **2024**, *52*, 72–76.
19. Wang, F.Y.; Huang, Z.X.; Liu, Y.; Li, Y. Novel cardanol-containing boron-modified phenolic resin composites: Non-isothermal curing kinetics, thermal properties, and ablation mechanism. *High Perform. Polym.* **2017**, *29*, 279–288.
20. Wang, S.J.; Jing, X.L.; Wang, Y.; Si, J. High char yield of aryl boron-containing phenolic resins: The effect of phenylboronic acid on the thermal stability and carbonization of phenolic resins. *Polym. Degrad. Stab.* **2014**, *99*, 1–11.
21. Ma, X. *Material Optimization and Friction and Wear Performance of Clamp-Type All-Carbon Ceramic Friction Sub-Materials*; Northwestern Polytechnical University: Xi'an, China, 2023.
22. Zhang, C.S.; Zhao, F.; Zhang, J.J.; Wang, X.Y.; Bao, G.L.; Bai, Y.B.; Li, T.J. Infrared spectroscopic study of nano-sized alumina. *J. Chem.* **1999**, *3*, 275–280.
23. Zhao, G.L.; Wang, S.J.; Ling, F.X.; Wu, H.; Cui, X.; Zhang, Y. Infrared spectroscopic study of the surface structure of γ-Alumina. *Contemp. Chem. Ind.* **2012**, *41*, 661–663+667.
24. Starczewski, L.; Szumniak, J. Mechanisms of transferring the matter in a friction process in a tribology system: Polymeric composite-metal. *Surf. Coat. Technol.* **1998**, *100*, 33–37. [CrossRef]
25. CSDN Library. Steel Heating Temperature Vs. Tolor Mapping. 2021. Available online: https://download.csdn.net/download/zgzmdr/56419228/ (accessed on 5 December 2021).
26. Dai, Y.J.; Bi, G.J. Application of low carbon steel quenching and strengthening process in chain drive industry. *Heat Treat. Technol. Equip.* **2024**, *45*, 37–41.

materials

MDPI

Article

Glass-Ceramics of the Lithium Aluminosilicate System Nucleated by TiO_2: The Role of Redox Conditions of Glass Melting in Phase Transformations and Properties

Stanislav Maltsev [1], Olga Dymshits [2,*], Irina Alekseeva [2], Anna Volokitina [2], Maksim Tenevich [3], Anastasia Bachina [3], Kirill Bogdanov [4], Svetlana Zapalova [2], Georgiy Shakhgildyan [5] and Aleksandr Zhilin [6]

1 Institute of Machinery, Materials, and Transport, Peter the Great St. Petersburg Polytechnic University, St. Petersburg 195251, Russia; riplgandeleon@mail.ru
2 S.I. Vavilov State Optical Institute, St. Petersburg 192171, Russia; vgolub@gmail.com (I.A.); anna.itmo@gmail.com (A.V.); zenii99@yandex.ru (S.Z.)
3 Ioffe Institute, St. Petersburg 194021, Russia; mtenevich@gmail.com (M.T.); a.k.bachina@yandex.ru (A.B.)
4 International Research and Education Center for Physics of Nanostructures, ITMO University, St. Petersburg 199034, Russia; kirw.bog@gmail.com
5 Department of Glass and Glass-Ceramics, Mendeleev University of Chemical Technology, Moscow 125047, Russia; georgiy.shahgildyan@gmail.com
6 D.V. Efremov Institute of Electrophysical Apparatus, St. Petersburg 189631, Russia; zhilin1311@yandex.ru
* Correspondence: vodym1959@gmail.com

Abstract: Lithium aluminosilicate glasses nucleated by TiO_2 are usually melted in oxidizing conditions. The reducing conditions of glass melting, which allow to obtain ions of variable valence in lower oxidation states, can influence the ability of titania to provide proper phase assemblage, structure and properties of lithium aluminosilicate glass-ceramics. The aim of this study is to reveal this influence. The model glass containing TiO_2 was melted with and without the addition of As_2O_3. Using heat treatments between 680 °C and 1300 °C, XRD, SEM and DSC data, Raman and absorption spectroscopy, transparent glass-ceramics based on nanocrystals of β-quartz solid solutions (ss) and/or γ-Al_2O_3 with spinel structure and opaque glass-ceramics based on nanocrystals of β-spodumene ss were obtained and characterized. Three-phase immiscibility develops during secondary heat treatments. Al_2TiO_5 crystallizes from aluminotitanate amorphous regions simultaneously with the appearance of β-quartz ss, while traces of anatase and then rutile appear at elevated temperatures. Phase assemblage and the sequence of phase transformations do not depend on the redox conditions of glass melting, while the rate of these transformations is significantly higher in glass melted without the addition of As_2O_3. Absorption in glass melted without the addition of As_2O_3 and the corresponding glass-ceramics originate from octahedrally coordinated Ti^{3+} ions and Ti^{3+}-Ti^{4+} pairs in glass and nanocrystals of γ-Al_2O_3, Al_2TiO_5 and β-quartz ss. Transparent glass-ceramics with a thermal expansion coefficient of ~0.3×10^{-6} K^{-1} were obtained from both glasses.

Keywords: redox conditions; titanium dioxide; nucleating agent; phase transformation; glass-ceramics; nanocrystals; Raman spectroscopy; absorption spectroscopy; coefficient of thermal expansion; X-ray diffraction analysis

1. Introduction

Transparent lithium aluminosilicate glass-ceramics combine the unique properties of transparency and a close-to-zero thermal expansion coefficient [1–3]. Developed by Stookey more than 60 years ago [4], glass-ceramics of the lithium aluminosilicate system still remain a subject of intensive studies [5–15]. Stookey found out [4] that titania, TiO_2, is an effective nucleating agent promoting crystallization of lithium aluminosilicate glasses of special compositions upon heat treatments. The role of titania as a nucleating agent of the lithium aluminosilicate system was studied by different structure-sensitive methods [16–22] and mechanisms controlling the formation of titania-doped glass-ceramics were suggested. It was demonstrated that they depend on the base glass composition, the nature and concentration of doping ions, and the amount of the nucleating agent. Being a polyvalent ion, titanium can be found in glasses in the form of Ti^{4+} and Ti^{3+} ions. Oxidation states of titanium ions depend on the glass melting temperature, oxygen partial pressure in the glass-melting furnace, the concentration of titanium ions and the glass composition [18,23–29]. Using titania as a nucleator, oxidizing agents are usually added to the glass batch to avoid unwanted coloration caused by traces of Ti^{3+} ions formed during the glass melting at high temperatures [1,2]. However, the reducing atmosphere of glass melting is required to obtain polyvalent ions in lower oxidation states. Therefore, it is worth knowing how the reducing atmosphere of glass melting influences phase transformations in glasses of the lithium aluminosilicate system nucleated by TiO_2 and properties of corresponding glass-ceramics.

There are studies of phase transformations in titania-containing glasses of magnesium aluminosilicate [18,29,30], fresnoite [31–33], and zinc aluminosilicate [34] systems melted under different redox conditions. In these studies, it was demonstrated that variation of the glass melting conditions influences kinetics of liquid–liquid phase separation and of crystallization of metastable phases, compositions and structures of titania-containing phases and generally does not influence crystallization of the stable equilibrium phases [18,29,34]. The aim of the present work is to study the effect of redox conditions of glass melting on the phase composition and properties of lithium aluminosilicate glass-ceramics nucleated by titania, which will help to develop transparent thermal shock resistant glass-ceramics of the lithium aluminosilicate system containing polyvalent ions in lower oxidation states important for advanced photonic applications [35].

2. Materials and Methods

2.1. Materials Preparation

Model glass with the composition 12 $Li_2O{\cdot}24\ Al_2O_3{\cdot}64\ SiO_2$ (mol%) [17,19] nucleated by 6 mol% TiO_2 added on top of the base composition was melted in oxidizing (with addition of 0.5 wt% As_2O_3) and neutral (without addition of arsenic trioxide) conditions in a laboratory electric furnace in crucibles made of quartz ceramics at 1580 °C for 4 h (h) with stirring and casted onto a metal plate. The raw materials were reagent grade oxides and lithium carbonate manufactured by Neva-Reactive Co, Saint-Petersburg, Russian Federation. The weight of the glass batch was 400 g. The glasses were annealed at 640 °C for 1 h and cooled with an annealing furnace to room temperature. Small portions of the glasses were quenched by pressing with a metal plate and used for the differential scanning calorimetry (DSC) study. Annealed glasses were transparent and different in color. The glass melted under neutral conditions, referred to as LAS, exhibited a brownish-grey color, while the LAS_{ox} glass, prepared under oxidizing conditions, displayed a light-yellow tint. The color difference is clearly seen in photographs of the as-cast glasses with a thickness of ca. 6 cm (see Figure 1).

The glasses were cut into pieces with a diamond saw and isothermally heat-treated at temperatures between 680 °C and 1200 °C by single and two-stage heat treatments for 6 h at each stage.

Figure 1. Photographs of as-casted glasses with thickness of ca. 6 mm: (**a**) LAS; (**b**) LAS_{ox}. Glass notations are shown in figure.

2.2. Materials Characterization

2.2.1. The Differential Scanning Calorimetry

The differential scanning calorimetry study was performed using a simultaneous thermal analyzer NETZSCH STA 449 F3 Jupiter, Netzsch Geratebau GmbH, Selb, Germany, with an Ar dynamic flow atmosphere. The temperature was increased from room temperature to 1300 °C at a heating rate of 10 °C·min^{-1}. The quenched glasses and those subjected to single stage holdings at 680 °C, 700 °C, and 720 °C for 6 h, with a weight of about 15–20 mg were employed. To determine the character of crystalline phases responsible for exothermic effects on the DSC curves, quenched and heat-treated glasses with a weight of ca. 100 mg were heated in the furnace of the thermal analyzer up to corresponding temperatures. The glass-ceramics obtained were taken out from the cooled furnace and studied by X-ray diffraction (XRD) analysis.

2.2.2. XRD Analysis

The XRD study of the powdered samples was performed using a Shimadzu XRD-6000 diffractometer, Shimadzu, Kyoto, Japan, with Cu Kα radiation and a Ni filter (λ = 1.5406 Å). The mean crystal sizes were calculated from the broadening of X-ray peaks according to Scherrer's equation:

$$D = K\lambda/\Delta(2\theta)\cos\theta, \quad (1)$$

where λ is the wavelength of X-ray radiation, θ is the diffraction angle, Δ(2θ) is the width of peak at half of its maximum, and K is the constant assumed to be 1 [36]. The mean size of γ-Al_2O_3 crystals with a spinel structure was estimated using the peak with Miller's indices *hkl* (440). The mean size of tieilite, Al_2TiO_5, crystals was determined from the peak with Miller's indices (020). The mean size of crystals of β-quartz ss was determined from the peak with Miller's indices (220). The mean size of crystals of β-spodumene ss was calculated using the peak with indices (102). An error of the calculation of the mean size of crystallites is about 5–10%.

The lattice parameter *a* of spinel nanocrystals was estimated from the position of the peak with *hkl* indices (440). The lattice parameters of Al_2TiO_5 were estimated using the peaks with indices (002), (020), (110), (023) and (200). The lattice parameters of β-quartz ss were calculated from the peaks with indices (110) and (211). The lattice parameters of β-spodumene ss were calculated from the peaks with *hkl* indices (111) and (102). An error in the estimation of the lattice parameters is ±0.003 Å.

2.2.3. Raman Spectroscopy

Unpolarized Raman spectra were obtained using a backscattering configuration with a confocal InVia Renishaw Raman microscope, Renishaw, Wotton-under-Edge, England. The setup included a Leica objective with a magnification of ×50 (N.A. = 0.75), a thermoelectrically cooled CCD camera, and an edge filter. For excitation, a laser with a wavelength of 488 nm (from an Ar+ ion source) was employed. The spectral resolution achieved during the measurements was 2 cm^{-1}. Each spectrum was obtained by averaging data from

10 acquisitions, each lasting 10 s. The samples were either transparent polished flat-parallel plates having a thickness of ~1 mm or opaque pieces of materials.

2.2.4. Scanning Electron Microscopy

The morphology of the initial and heat-treated glasses was obtained by scanning electron microscopy (SEM) using a Tescan Vega 3 SBH microscope, Tescan, Brno, Czech Republic. For the study, the surfaces of the same transparent polished plates and opaque bulk samples, which were used for measuring Raman spectra, were preliminarily cleaned in isopropyl alcohol and benzene, etched in a solution of HF for about 2 s and washed in distilled water. ImageJ software version 1.51j8 [37] was used to calculate the particle sizes.

2.2.5. Absorption Spectroscopy

A Shimadzu UV-3600 spectrophotometer, Shimadzu, Japan, was employed for recording absorption spectra within the wavelength range from 250 to 3300 nm using the same polished ~1 mm thick flat-parallel plates.

2.2.6. The Linear Coefficient of Thermal Expansion

The thermal expansion coefficients of the glasses and glass-ceramics were estimated using a Linseis L 75 vs. 1000 dilatometer, Linseis Messgeräte GmbH, Selb, Germany. The samples were rods with a length ranging from 30 to 50 mm and flat-parallel planes with sections measuring ~5 × 5 mm. The measurements were performed at room temperature up to 320 °C with a heating rate of 5 °C·min^{-1}.

3. Results

The polished plates of transparent glasses and glass-ceramics and the pieces of opaque glass-ceramics are shown in Figure 2. Note that the thin polished plate of the glass made from the brownish-grey as-cast LAS glass has a light grey coloration because of a six times smaller thickness. The initial glasses, heat-treated at 680 °C for 6 h and by two-stage heat treatments with the second hold at temperatures between 720 °C and 900 °C, are transparent. The LAS glass-ceramics obtained at 950 °C and at 1000 °C are also transparent, while LAS_{ox} glass-ceramics obtained at these temperatures are translucent. Both glass-ceramics obtained at 1100 °C are translucent, and those prepared at 1200 °C are opaque. Therefore, glasses melted under neutral conditions lose transparency at higher secondary heat treatment temperatures than glasses melted under oxidizing conditions. The opaque LAS glass-ceramic prepared at 1200 °C is grey colored and has a white surface layer, see Figure 2a.

3.1. Study by Differential Scanning Calorimetry

Figure 3 shows the DSC scans of the quenched LAS and LAS_{ox} glasses, as well as these glasses heat-treated at 680 °C, 700 °C and 720 °C for 6 h. The shapes of the corresponding DCS curves of LAS and LAS_{ox} glasses are nearly similar, being different in the values of characteristic crystallization temperatures, see Figure 3 and Table 1. The DSC curve of the quenched LAS_{ox} glass shows three peaks with markedly different intensities. The first one is a broad peak of a low intensity with T_{max} = 802 °C, the second one is a narrow and intense peak with crystallization onset temperature T_{on} = 851 °C, and crystallization maximum temperature T_{max} = 874 °C. The third one is a weak exothermic peak with T_{max} = 1186 °C. The first peak is usually assigned to nucleation of the main crystalline phase, the second one to the crystallization of the lithium aluminosilicate solid solution (ss) with β-quarts structure (β-quarts ss), and the third one to the transformation of β-quartz ss into β-spodumene ss [38,39]. The DSC curve of the LAS glass exhibits a plateau instead of the first peak, see Figure 3 and Table 1. All characteristic crystallization temperatures of the

LAS glass are lower than those for the LAS_{ox} glass, while the glass transition temperatures of both glasses are similar, see Table 1. Therefore, the redox conditions of glass melting affect crystallization temperatures and do not affect the transition temperature of glasses.

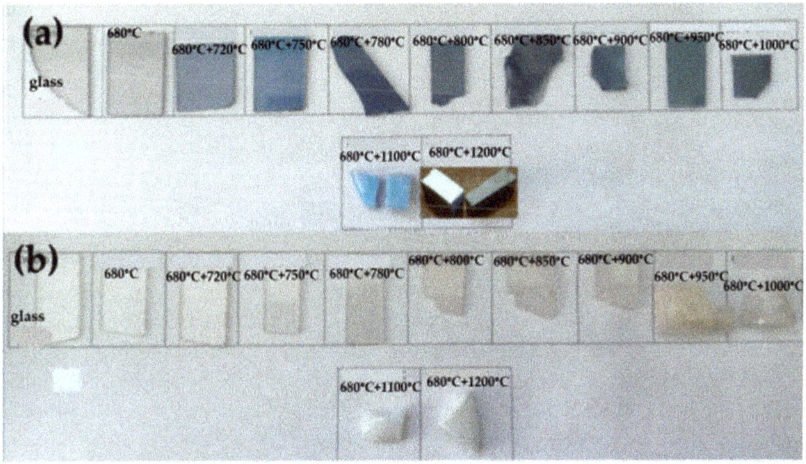

Figure 2. Photographs of samples of the precursor glasses, glasses heat-treated at 680 °C and glass-ceramics prepared by heat treatments with the first stage at 680 °C and the second stage between 720 °C and 1200 °C: (**a**) LAS; (**b**) LAS_{ox}. The regimes of heat treatments are shown in figure. Heat-treatment duration at each stage is 6 h. The thickness of transparent polished plates is 1 mm.

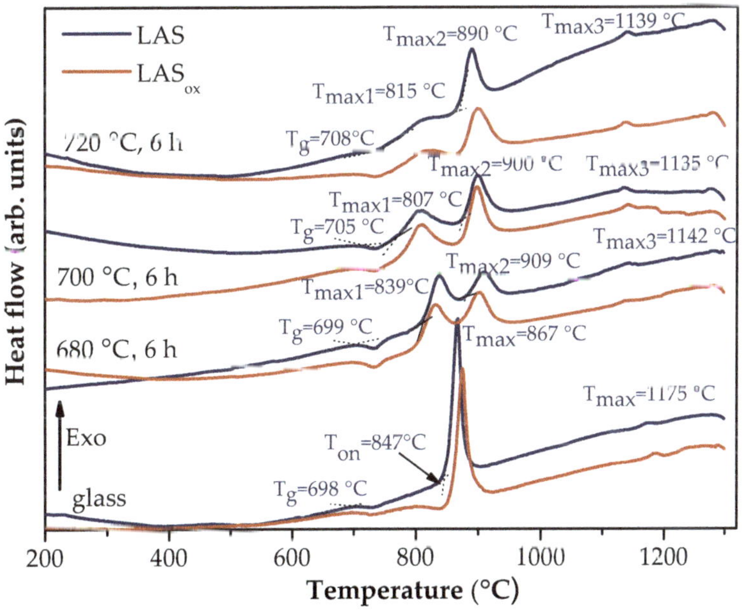

Figure 3. DSC signals as time functions for quenched and heat-treated LAS and LAS_{ox} glasses. T_g denotes the glass transition temperature, T_{on} is the onset crystallization temperature, T_{max} is the maximum crystallization temperature. The curves are shifted for the convenience of observation.

Table 1. Characteristic temperatures of initial quenched and heat-treated glasses shown in DSC curves.

Heat-Treatment Regime	T_g, °C	T_{on1}, °C	T_{max1}, °C	T_{on2}, °C	T_{max2}, °C	T_{max3}, °C
			The LAS glass			
quenched	698	-	-	847	867	1175
680 °C, 6 h	699	810	839	882	909	1142
700 °C, 6 h	705	764	807	878	900	1135
720 °C, 6 h	708	765	815	871	890	1139
			The LAS_{ox} glass			
quenched	699	-	802	851	874	1186
680 °C, 6 h	695	799	831	876	901	1136
700 °C, 6 h	704	779	809	879	899	1141
720 °C, 6 h	708	755	828	876	900	1136

Preliminary heat treatments lead to a drastic change in the appearance of DSC curves, see Figure 3. The character of this change for LAS and LAS_{ox} samples is the same, differing in the temperatures of thermal effects. Instead of a single narrow intense peak, two broad exothermic peaks of lower intensities appear. The characteristic temperatures for both glasses preliminarily heat-treated at 680 °C, 700 °C, and 720 °C are shown in Figure 3 and Table 1. T_{max} of the second broad exothermic peak is 35–40 °C higher than this temperature for the initial quenched glass. With the increase in the temperature of preliminary heat treatment, there is a redistribution of intensities of these two peaks in favor of the second one, see Figure 3. T_{max} of the second peak and T_g gradually increase, see Table 1.

The reason for this behavior will be explained below while discussing the phase composition of the samples. The characteristic T_g, T_{on} and T_{max} temperatures of glasses preliminary heat-treated at 680 °C are similar, being somewhat higher for the LAS glass than for the corresponding LAS_{ox} samples.

The XRD pattern of the glass-ceramic obtained by heating the quenched LAS glass in the furnace of the DSC instrument up to the temperature of the sharp exothermic peak at 880 °C, see Figure 4a, testifies to the crystallization of β-quarts ss, traces of tieilite, Al_2TiO_5, and spinel, see Figure 4b. On the XRD pattern of the sample heated up to the temperature of the third exothermic effect at 1175 °C, additional diffraction peaks appear as compared with the pattern of the sample heated up to 880 °C, manifesting in a transformation of the β-quarts ss into the β-spodumene ss while spinel traces disappeared. Thus, the small high-temperature peak evolving at ~1175 °C is due to crystallization of β-spodumene ss, see Figure 4b.

Figure 5a shows the DSC signals of the LAS glass preliminarily heat-treated at 680 °C for 6 h with stops at temperatures of 787 °C, 838 °C, 858 °C, and 935 °C. The corresponding XRD patterns are presented in Figure 5b. The material obtained by the heat treatment at 680 °C, 6 h + 787 °C, 0 h is X-ray amorphous. The sample obtained by the heat treatment at 680 °C, 6 h + 838 °C, 0 h contains nanocrystals of γ-Al_2O_3 with a spinel structure and traces of β-quarts ss. It is speculated that this exothermic peak is caused by the crystallization of spinel, while traces of β-quarts ss appear during the cooling of the sample in the furnace, because the samples heated to the specified temperature could not be immediately removed from the hot furnace of the DSC instrument. The XRD pattern of the sample obtained by the heat treatment at 680 °C, 6 h + 858 °C, 0 h, corresponding to the end of the first exothermic peak and the beginning of the second one, contains peaks of nanocrystals of γ-Al_2O_3, Al_2TiO_5 and β-quarts ss. The XRD pattern of glass-ceramic obtained by the heat treatment at 680 °C, 6 h + 935 °C, 0 h proves the crystallization of β-quarts ss and Al_2TiO_5. The presence of γ-Al_2O_3 cannot be ruled out on this XRD pattern because the most intense peak of γ-Al_2O_3 overlaps the peaks of β-quarts ss and is therefore difficult to detect.

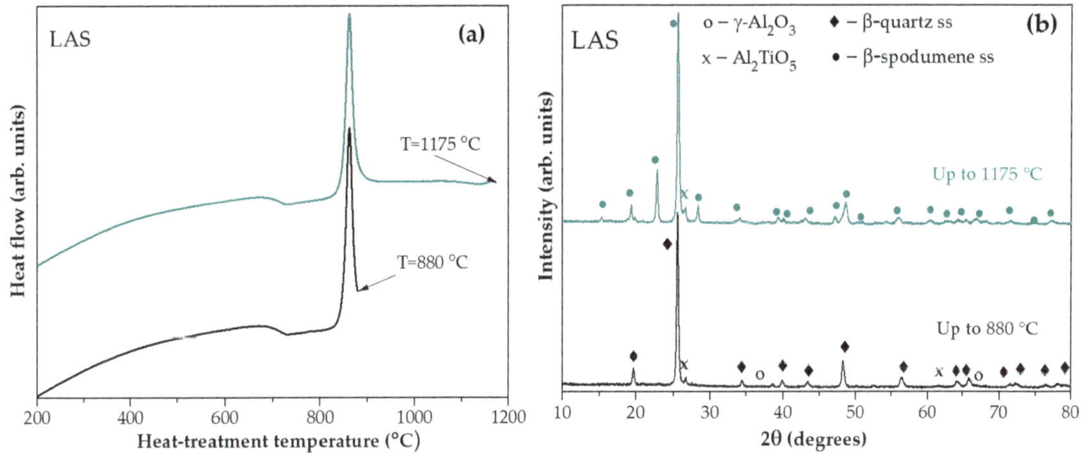

Figure 4. (**a**) DSC signals as time functions for the quenched LAS glass with stops at temperatures of 880 °C and 1175 °C, (**b**) XRD patterns of thus obtained LAS glass-ceramics. The curves are shifted for the convenience of observation.

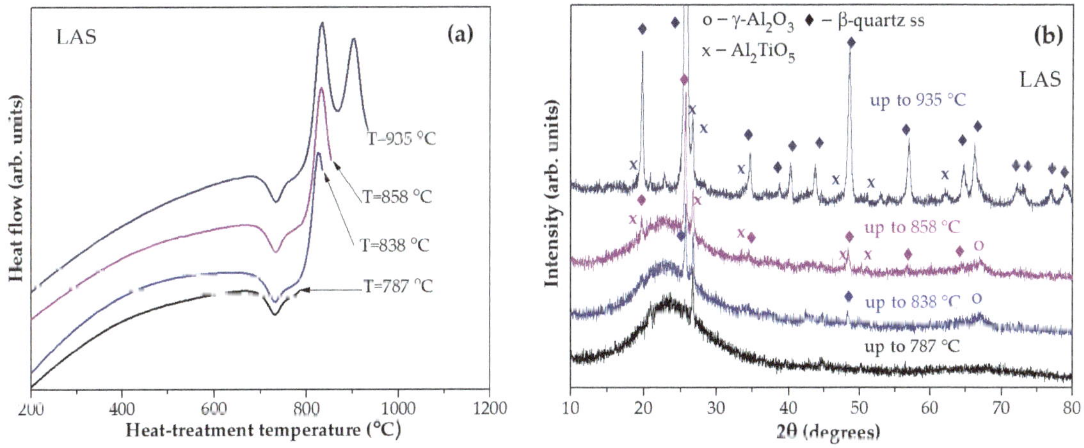

Figure 5. (**a**) DSC signals as time functions for the LAS glass preliminary heated at 680 °C for 6 h with stops at temperatures of 787 °C, 838 °C, 858 °C, and 935 °C; (**b**) XRD patterns of thus obtained LAS glass-ceramics. The curves are shifted for the convenience of observation.

The XRD pattern of the glass preliminarily heat-treated at 680 °C for 6 h and heated in the DSC furnace up to 858 °C is significantly different from the XRD pattern of the quenched glass heated in the DSC furnace up to 880 °C. The XRD pattern of the glass preliminarily heat-treated at 680 °C for 6 h and heated in the DSC furnace up to 935 °C is similar to the XRD pattern of the quenched glass heated in the DSC furnace up to 880 °C, see Figure 6. Thus, the first peak on the DSC curve of the preliminary heat-treated sample is associated with the crystallization of the phase with the spinel structure, and the second peak is caused by the crystallization of β-quartz ss and Al_2TiO_5. On the DSC curve of the quenched sample there is no exothermic peak associated with the crystallization of the phase with spinel structure. Therefore, preliminary heat treatments provoke spinel crystallization and increase the crystallization temperature of β-quartz ss and Al_2TiO_5.

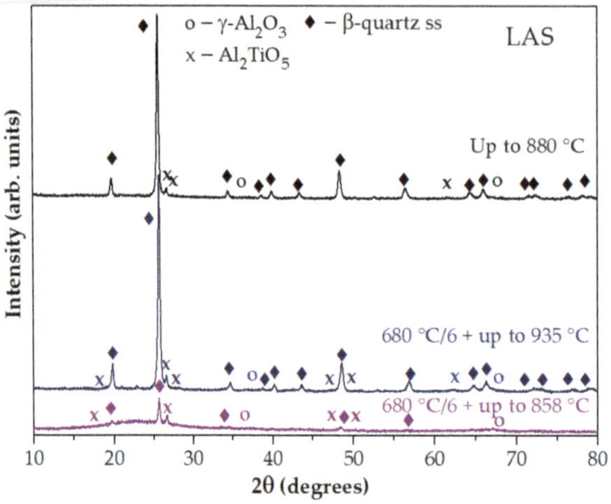

Figure 6. XRD patterns of the LAS quenched glass, and the glass heat-treated at 680 °C for 6 h and then heated in the furnace of the DSC instrument. The heat-treatment schedules are shown in the figure. The patterns are shifted for the convenience of observation.

3.2. XRD Study

Figure 7a,b shows XRD patterns of the LAS and LAS_{ox} glasses and glasses of the same compositions, which underwent heat treatment at 680 °C and two-stage heat treatments with a temperature at the second stage between 720 °C and 1200 °C. The holding time at each stage was 6 h. The figures allow the following of the formation of the main crystalline phase, β-quartz ss, at temperatures between 780 °C and 1000 °C, and crystallization of β-spodumene ss at temperatures between 1100 °C and 1200 °C. Figure 7c,d gives a closer look at the XRD patterns of the glasses and glass-ceramics fabricated by heat treatments at the second stage up to 800 °C. Initial LAS and LAS_{ox} glasses are X-ray amorphous with a maximum amorphous halo located at 2θ = 23.1° and 23.3°, respectively. Glasses heat-treated at the nucleation stage of 680 °C for 6 h are X-ray amorphous, and their XRD patterns are similar to those of glass precursors. The two-stage heat treatment with the second stage at 720 °C for 6 h leads to crystallization of a small fraction of crystallites with spinel structure, manifested by an appearance of a broad peak with Miller indices *hkl* (440) at 2θ ≅ 66.8°.

Its low intensity prevents estimation of the lattice parameter and the size of the spinel crystallites. The position of the amorphous halo shifts to a smaller angle 2θ = 22.8° for both samples. After the second stage heat treatment at 750 °C for 6 h, spinel peaks become more pronounced, which indicates an increase in the volume fraction of this phase. The lattice parameters of spinel nanocrystals are a_{norm} = 7.918 Å and a_{ox} = 7.916 Å, and sizes are D_{norm} = 4.5 nm and D_{ox} = 7.3 nm, respectively. Traces of β-quartz ss are also seen on the patterns. The appearance of spinel and traces of β-quartz ss cause a further change in the composition of the residual glass phase, which is manifested by a shift in the maximum of the amorphous halo to smaller angles, indicating that the residual glass becomes enriched in silica. After heat treatments at the second stage at 780 °C and 800 °C, the β-quartz ss becomes the predominant crystalline phase, while spinel and tieilite, Al_2TiO_5, nanocrystals are also found in the XRD patterns. The position of the amorphous halo of residual glass is shifted to 2θ = 22.6°and then to 2θ = 22.5° with increasing heat-treatment temperature, see Figure 7c,d.

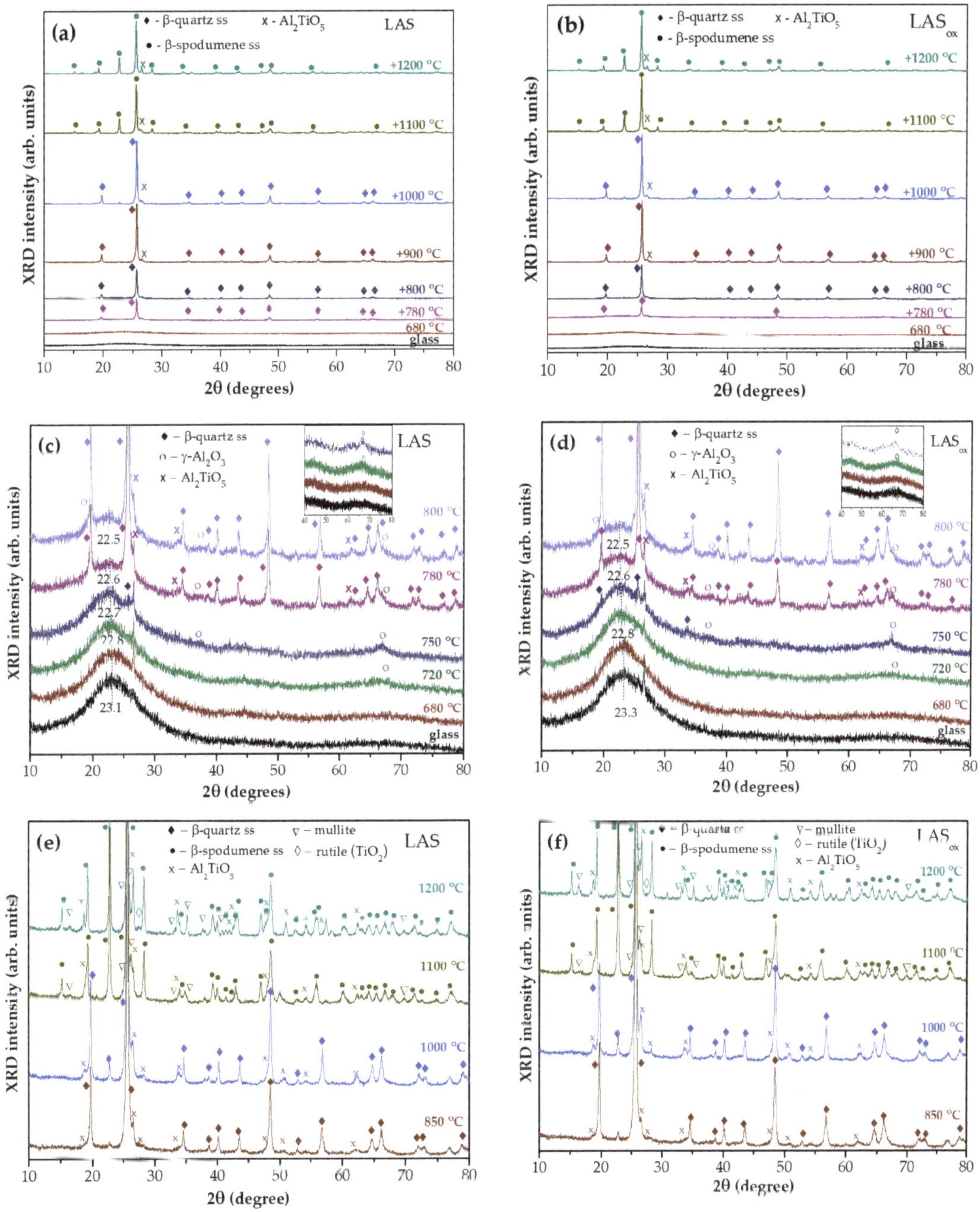

Figure 7. (**a**,**b**) XRD patterns of LAS and LAS_{ox} glasses and glass-ceramics: (**a**) the LAS; (**b**) the LAS_{ox}; (**c**,**d**) a closer look on XRD patterns of glasses and glass-ceramics obtained by heat treatments at the second stage up to 800 °C: (**c**) the LAS; (**d**) the LAS_{ox}; (**e**,**f**) a closer look on XRD patterns of glass-ceramics obtained by heat treatments at the second stage between 850 °C and 1200 °C: (**e**) the LAS; (**f**) the LAS_{ox}. The heat-treatment temperature at the first stage is 680 °C, duration of each stage is 6 h. The patterns are shifted for the convenience of observation.

The lattice parameters and mean crystallite sizes of spinel obtained at different heat-treatment temperatures are presented in Table 2. It is believed that spinel has the composition and structure of γ-Al_2O_3. The lattice parameter *a* of the unit cell of the cubic modification of Al_2O_3 usually takes the value *a* = 7.900–7.908 Å [40]. With a slight oxygen deficiency, the parameter *a* of γ-Al_2O_3 becomes equal to 7.911 Å (ICDD PDF card #79-1558) and 7.914 Å (ICDD PDF card 79-1557) [41]. The lattice parameter *a* of γ-Al_2O_3 crystallites in glass-ceramics changes from 7.915 Å to 7.925 Å and increases with increasing the heat-treatment temperature. The difference in the parameter *a* of γ-Al_2O_3 nanocrystals in LAS and LAS_{ox} glass-ceramics is very small. Nevertheless, the lattice parameter *a* of γ-Al_2O_3 in LAS glass-ceramics is slightly higher than in LAS_{ox} ones. The sizes of spinel crystallites increase with increasing the heat-treatment temperature ranging from 4.5 to 14.0 nm, see Table 2.

Table 2. Lattice parameters and mean crystallite sizes of γ-Al_2O_3, β-quartz ss and β-spodumene ss in LAS and LAS_{ox} glass-ceramics.

Heat-Treatment Regime	γ-Al_2O_3		β-Quartz ss			β-Spodumene ss		
	a, Å	D, nm	*a*, Å	*c*, Å	D, nm	*a*, Å	*c*, Å	D, nm
The LAS glass								
680 °C, 6 h + 720° C, 6 h								
680 °C, 6 h + 750 °C, 6 h	7.918	4.5						
680 °C, 6 h + 780° C, 6 h	7.921	8.0	5.207	5.336	26			
680 °C, 6 h + 800 °C, 6 h	7.925	14.0	5.196	5.350	25			
680 °C, 6 h + 850 °C, 6 h			5.189	5.370	22			
680 °C, 6 h + 1000 °C, 6 h			5.186	5.362	28			
680 °C, 6 h + 1100 °C, 6 h						7.552	9.145	45
680 °C, 6 h + 1200 °C, 6 h						7.541	9.143	45
The LAS_{ox} glass								
680 °C, 6 h + 720° C, 6 h								
680 °C, 6 h + 750 °C, 6 h	7.916	7.3			28			
680 °C, 6 h + 780° C, 6 h	7.915	9.4	5.207	5.319	26			
680 °C, 6 h + 800 °C, 6 h	7.921	10.3	5.197	5.337	26			
680 °C, 6 h + 850 °C, 6 h			5.192	5.365	21			
680 °C, 6 h + 1000 °C, 6 h			5.187	5.368	27			
680 °C, 6 h + 1100 °C, 6 h						7.550	9.153	36
680 °C, 6 h + 1200 °C, 6 h						7.551	9.150	36

Figure 8a shows that in spite of the same phase assemblage of glass-ceramics obtained by two-stage heat treatment at 680 °C, 6 h + 780 °C, 6 h, the crystallinity fractions in the LAS glass-ceramic is much higher than in the LAS_{ox} one. This difference is levelled out by the heat treatment at 680 °C, 6 h + 800 °C, 6 h, see Figure 8b.

Figure 7e,f show the XRD patterns of glass-ceramics obtained by two-stage heat treatments with a second-stage temperature between 850 °C and 1200 °C. After the heat treatment at 850 °C the crystallinity fraction increases to such extent that the amorphous halo disappears. Volume fractions of β-quartz ss and Al_2TiO_5 gradually increase. β-spodumene ss crystallizes at the expense of β-quartz ss at 1100 °C. Mullite (ICDD PDF card # 79-1458) appears at the same heat-treatment temperature. The lattice parameters and mean crystallite sizes of β-quartz ss and β-spodumene ss are presented in Table 2. Their variation with heat-treatment temperature is similar for the LAS and the LAS_{ox} glasses.

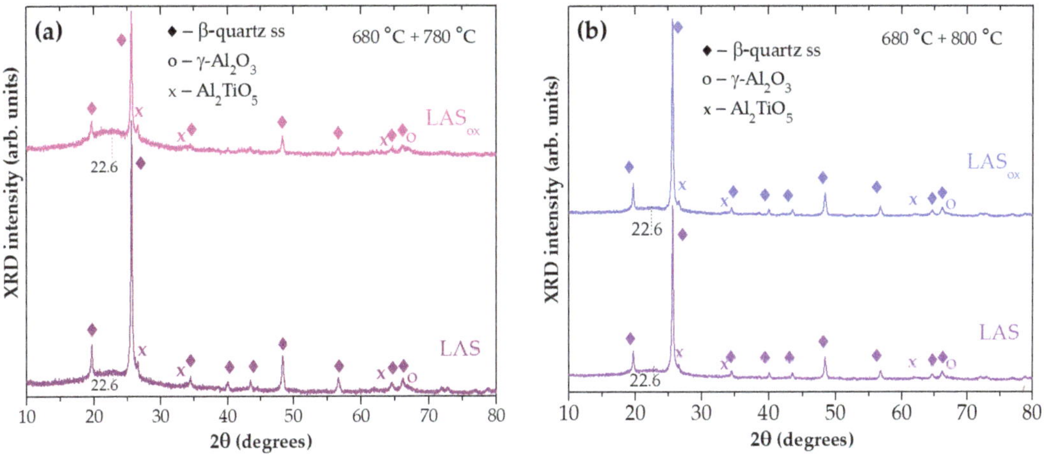

Figure 8. XRD patterns of the LAS and the LAS_{ox} glass-ceramics obtained by two-stage heat treatments at: (**a**) 680 °C, 6 h + 780° C, 6 h; (**b**) 680 °C, 6 h + 800 °C, 6 h. The patterns are shifted for the convenience of observation.

As was mentioned above, the appearance of the surface and the volume of the LAS glass-ceramics obtained by the heat treatment at 1200 °C are different. The surface is a dense, white-colored layer while the volume is grey colored, see Figure 2. Traces of rutile (ICDD PDF card #78-1509) are found on the XRD patterns taken from the grey part of the LAS glass-ceramic. The appearance of rutile is accompanied by some decrease in the tieilite fraction. We failed to find a difference in the phase assemblages of the surface and volume of LAS_{ox} glass-ceramic obtained by the same heat treatment. The XRD pattern shows a tiny fraction of rutile in this sample.

Let us discuss the features of crystallization of tieilite, Al_2TiO_5. As mentioned above, the onset of the crystallization of β-quartz solid solution in glasses at a temperature of 780 °C is accompanied by the appearance of a small fraction of Al_2TiO_5 crystallites with mean sizes of ~6 and 8 nm for LAS and LAS_{ox} glass-ceramics, respectively, see Table 3. Increasing the heat-treatment temperature up to 1200 °C leads to an increase in the size of Al_2TiO_5 crystallites up to ~40 nm, a slow increase in their fraction and a change in their lattice parameters, see Table 3. The values of the lattice parameter do not change with the heat-treatment temperature within the measurement error and are independent of the redox conditions of glass melting. Lattice parameters b, c and the volume increase with the heat-treatment temperature. An analysis of the evolution of lattice parameters shows that the lattice of a Al_2TiO_5 unit cell, which has the shape of a rectangular parallelepiped elongated in [010] and [001] directions, becomes even more elongated in the same directions under the influence of high temperature. The lattice parameters of tieilite in LAS and LAS_{ox} glass-ceramics are similar. The character of their variation with temperature is also similar. The redox conditions of glass melting do not significantly affect the structural transformations in tieilite crystallites. A pronounced change in lattice parameters after the heat treatment at 1200 °C found in the present study can be a prerequisite to decomposition of tieilite with exsolution of rutile [42]. Indeed, as mentioned above, traces of rutile are found in LAS_{ox} and in the volume of the LAS glass-ceramics obtained by the heat treatment at 1200 °C. The comparison of XRD patterns of the LAS and the LAS_{ox} glass-ceramics obtained by the heat treatment at 1200 °C demonstrates that LAS_{ox} glass-ceramic contains a smaller fraction of rutile and a higher fraction of mulite than the LAS glass-ceramic, compare Figure 7e,f.

Table 3. Lattice parameters and mean crystallite sizes of Al_2TiO_5 in LAS and LAS_{ox} glass-ceramics.

Heat-Treatment Regime	Al_2TiO_5									
	The LAS Glass					The LAS_{ox} Glass				
	a, Å	*b*, Å	*c*, Å	V, Å^3	D, nm	*a*, Å	*b*, Å	*c*, Å	V, Å^3	D, nm
680 °C, 6 h + 780° C, 6 h	3.594	9.356	9.557	321.3	6.2	3.593	9.541	9.577	321.4	8.0
680 °C, 6 h + 800 °C, 6 h	3.595	9.365	9.564	322.0	7.1	3.594	9.358	9.577	322.1	8.7
680 °C, 6 h + 850 °C, 6 h	3.592	9.378	9.560	322.0	8.3	3.595	9.385	9.589	323.5	10.2
680 °C, 6 h + 900 °C, 6 h	3.596	9.373	9.570	322.6	9.0	3.594	9.418	9.619	325.6	12.0
680 °C, 6 h + 950 °C, 6 h	3.597	9.454	9.590	326.1	9.4	3.594	9.421	9.628	326.0	12.0
680 °C, 6 h + 1000 °C, 6 h	3.598	9.456	9.624	327.4	12.2	3.594	9.457	9.633	327.4	13.7
680 °C, 6 h + 1100 °C, 6 h	3.596	9.456	9.663	328.6	18.8	3.595	9.460	9.677	329.1	19.4
680 °C, 6 h + 1200 °C, 6 h	3.591	9.488	9.716	331.0	44.1	3.592	9.489	9.708	330.9	42.0

The sequence of phase transformations revealed by XRD study is similar for both glasses: precursor glasses and glasses subjected to the nucleation treatment are X-ray amorphous; nanocrystals of γ-Al_2O_3 with spinel structure and sizes ranging from 4.5 nm to 14 nm evolve during heat treatments in the second stage in the temperature range from 720 °C to 800 °C; β-quartz ss and the crystalline phase of the nucleating agent, tieilite, appear additionally to spinel during the heat treatment at 780 °C; β-quartz ss are obtained by heating in the interval between 780 °C and 1000 °C; the glass-ceramic prepared at 1100 °C contains β-spodumene ss instead of β-quartz ss and traces of mullite; β-spodumene ss is the main crystalline phase of glass-ceramics produced in the temperature range between 1100 °C and 1200 °C; tieilite nanocrystals with a mean size ranging from ~6 nm to ~45 nm are found in glass-ceramics prepared in the temperature interval from 800 °C to 1200 °C; during the heat treatment at 1200 °C, the rutile crystallites appear in the phase assemblage of glass-ceramics.

In spite of the same phase assemblage of glass-ceramics prepared from glasses melted in different redox conditions, the kinetics of phase transformations and spinel lattice parameters are slightly different. The spinel lattice parameters in the LAS glass-ceramics are larger than in the LAS_{ox} glass-ceramics. The crystallinity fractions in the LAS glass-ceramic obtained by two-stage heat treatment at 680 °C, 6 h + 780 °C, 6 h is much higher than in the LAS_{ox} one. However, this difference is levelled out by the heat treatment at 680 °C, 6 h + 800° C, 6 h. The rutile crystallinity fraction in the LAS glass-ceramics is larger than in the LAS_{ox} one, while the mullite fraction is smaller.

3.3. *Study by Raman Spectroscopy*

Figure 9a,b show Raman spectra of the initial and heat-treated LAS glass. Raman spectra of initial LAS and LAS_{ox} glasses are similar and contain broad bands with maxima at 482 cm^{-1}, 800 cm^{-1}, 910 cm^{-1} and ~1015 cm^{-1}. The wing of the latter band extends to 1200 cm^{-1}. A similar spectrum was obtained in ref. [19] for the glass of the same composition nucleated by 7 mol% TiO_2 and melted in oxidizing conditions. The bands at 482 cm^{-1}, 800 cm^{-1}, and ~1000–1200 cm^{-1} are due to vibrations of bonds in the tetrahedrons of the aluminosilicate network in the glass structure [43], and the band at ~910 cm^{-1} is due to vibrations of $[TiO_4]$ tetrahedra embedded in this network [44]. After the heat treatment at the nucleation stage at a temperature of 680 °C, minor changes are noticed in the Raman spectrum, see Figure 9a,b. The high-frequency band slightly broadens and the band with a maximum at 800 cm^{-1} somewhat increases in intensity compared to the band with a maximum at 910 cm^{-1}, which indicates the development of liquid–liquid phase separation of the precursor glass [19]. The position of the band at 482 cm^{-1} is not influenced by this heat treatment.

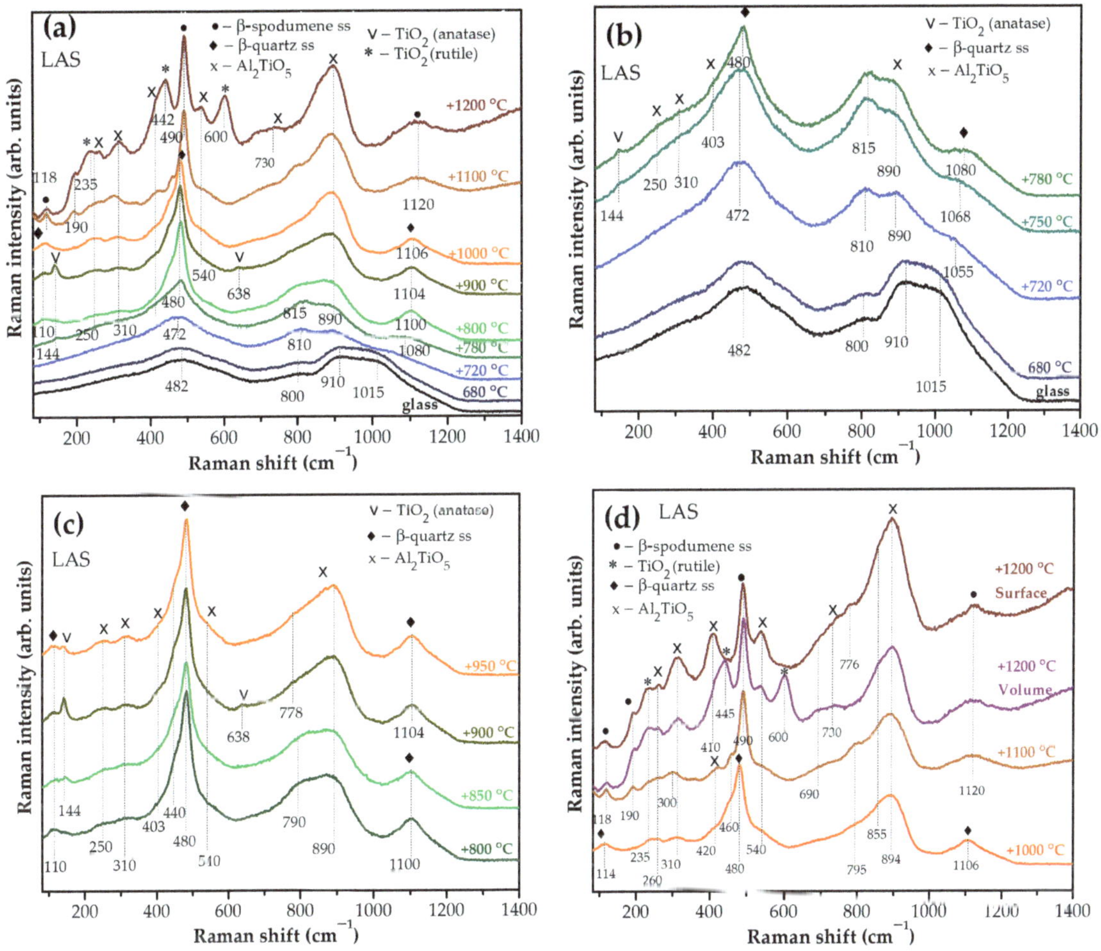

Figure 9. Raman spectra of the LAS glass and glass-ceramics obtained by single and two-stage heat treatments: (**a**) from the glass to 1200 °C; (**b**) from the glass up to 780 °C; (**c**) from 800 °C up to 950 °C; (**d**) from 1000 °C up to 1200 °C. The time of heat treatment at each hold is 6 h. The spectra are shifted for the convenience of observation.

After the two-stage heat treatment with a temperature of 720 °C at the second stage, significant changes are observed in the Raman spectrum, see Figure 9a,b. The bands with maxima at 910 cm^{-1}, 800 cm^{-1}, and 482 cm^{-1} move their positions to 890 cm^{-1}, 810 cm^{-1}, and 472 cm^{-1}, respectively. Instead of the band with a maximum at ca. 1015 cm^{-1}, a band at 1055 cm^{-1} appears. The intensity of the band at 472 cm^{-1} increases as compared with the similar band in spectra of precursor glass and glass heat treated at the nucleation stage. The intensity of the band at ~810 cm^{-1} increases relatively to the band at ca. 890 cm^{-1}. The increase in the intensity of this band at the expense of the band with a maximum at ca. 890 cm^{-1} is due to the superposition of vibrations of the $[TiO_5]$ and $[TiO_6]$ groups in amorphous aluminotitanate regions on a weak band in the 800 cm^{-1} region, corresponding to vibrations of the tetrahedrons of the aluminosilicate network [17,19,20]. These changes are associated with a further development of the liquid–liquid phase separation with the formation of aluminotitanate amorphous regions [19,20] and a change in the composition of the aluminosilicate glass network, which is in accordance with a change in the position of the amorphous halo in the corresponding XRD pattern, see Figure 7c. Crystallization of γ-Al_2O_3 with spinel structure revealed by XRD analysis does not show itself in the Raman spectrum of this glass-ceramic. Running ahead, one will say that spectral features of γ-Al_2O_3, were not detected in Raman spectra of samples obtained by heat treatments between 750 °C and 800 °C as well, in spite of the fact that XRD analysis reliably identified them. The reason is that bond vibrations in aluminate spinel crystals are very weak compared to bond vibrations in titanium-containing compounds. According to refs [45,46], γ-Al_2O_3 exhibits narrow peaks with maxima at 315, 410, 520, 713 and 835 cm^{-1} in the Raman spectrum of γ-Al_2O_3 corresponding to vibrations of the Al–O bond in tetrahedral structural units of AlO_4 [45]. Based on calculations presented in ref. [47], the strongest Raman peak for γ-Al_2O_3 with a spinel structure is located at ~401 cm^{-1}. There are also several bands of medium intensities ranging from 100 cm^{-1} to 900 cm^{-1} [47]. In Raman spectra presented in Figure 9b there is a band with a maximum at 403 cm^{-1}, which could be the spectroscopic sign of γ-Al_2O_3. However, this band can be also caused by vibrations in tieilite crystals, because this band appears simultaneously with other bands at 250 cm^{-1}, 310 cm^{-1}, and 890 cm^{-1}, characteristic of tieilite, which crystallizes at a larger temperature interval of heat treatments than γ-Al_2O_3 (see below). It is worth mentioning that the spectroscopic signs of crystals with a spinel structure were found only in Raman spectra of spinel-based glass-ceramics of magnesium [48] and zinc aluminosilicate systems [34], where the spinel crystallinity fraction was significantly higher than in the present case.

In the Raman spectrum of the glass-ceramic obtained by the heat treatment at the second stage at 750 °C, the bands with maxima at 472 cm^{-1} and 815 cm^{-1} are enhanced, positions of the band maxima change from 810 cm^{-1} to 815 cm^{-1} and from 1055 cm^{-1} to 1068 cm^{-1}, and a very weak band appears at ~144 cm^{-1}. The latter peak can be attributed to the most intense vibration in the Raman spectrum of the metastable modification of TiO_2, anatase [17].

In the Raman spectrum of the sample obtained by the heat treatment at the second stage at a temperature of 780 °C, the band at ~144 cm^{-1}, caused by vibrations in anatase nanocrystals, narrows, the band at 480 cm^{-1} narrows and intensifies, and the band at ~1080 cm^{-1} appears. The last two bands are related to vibrations in nanocrystals of β-quartz ss [17,19,49–51]. As mentioned above, in the spectrum of these glass-ceramics there is a number of bands with maxima at 250 cm^{-1}, 310 cm^{-1}, 403 cm^{-1}, 540 cm^{-1}, and 890 cm^{-1}. They appear simultaneously, and their intensities increase continuously with the increasing temperature of heat treatment. These bands belong to vibrations in tieilite nanocrystals [17,19], which is in accordance with the XRD data. The Raman spectrum of glass-ceramic obtained by heat treatment with a temperature of 900 °C at the second stage shows two anatase bands at ~144 cm^{-1} and ~638 cm^{-1} manifesting the maximum anatase crystallinity fraction achieved by this heat treatment, see Figure 9a,c. After increasing the temperature of the second hold to 1000 °C, the bands related to anatase disappear. Note that the XRD peaks of anatase coincide in position with the main peaks of β-quartz ss and its small amount cannot be detected by the XRD data [17,19].

As the temperature at the second stage of heat treatment increases, the intensities of the bands at ~460 cm^{-1} and ~1080 cm^{-1} increase, and their peaks become narrower, see Figure 9c, which indicates the development of β-quartz ss crystallization. Their positions constantly move to longer wave numbers manifesting enrichment of these crystals with silica [50]. After the second stage heat treatment at 1100 °C, these bands shift to 490 cm^{-1} and 1120 cm^{-1}, which can be interpreted as recrystallization of β-quartz ss into β-spodumene ss [17,19]. Starting from the heat treatment at 800 °C, a weak band appears at 110 cm^{-1} simultaneously with increasing intensity and narrowing of the bands attributed to β-quartz ss. This band can correspond to external vibrations in nanocrystals of β-quartz ss [51]. In

Raman spectra of glass-ceramics obtained by heat treatments at 1100 °C and at 1200 °C, weak bands at ~118 cm^{-1} and ~190 cm^{-1} were observed that can be attributed to external vibrations in nanocrystals of β-spodumene ss [49], see Figure 9d.

Figure 9d shows the bands at ~260 cm^{-1}, ~445 cm^{-1} and ~600 cm^{-1}, which are related to crystallization of rutile [17] at the expense of tieilite in the bulk of glass-ceramic obtained by the heat treatment at a second stage at 1200 °C. The signs of rutile crystals were not found in the spectrum from the white surface of this sample, which is in accordance with XRD data. XRD patterns of glass-ceramics obtained by heat treatments at 1100 °C and 1200 °C contain small mullite fractions, see Figure 7e,f. Raman spectra of synthetic mullites of different compositions have similar spectra with the strongest bands located at 415 cm^{-1}, 600 cm^{-1}, and 980 cm^{-1} [52]. They can be hidden in the contours of the corresponding broad bands in Raman spectra, see Figure 10f.

Raman spectra of glass-ceramics obtained by the same heat treatments that differ most strongly from each other according to Raman spectroscopy data are presented in Figure 10a–f. Raman spectra of glass precursors, as well as the spectra of glasses heat-treated at 680 °C for 6 h are similar to each other. The spectra of glass-ceramics obtained by the two-stage heat treatment at 680 °C and at 720 °C for 6 h differ from each other by the ratio of band intensities in the region of high wave numbers, see Figure 10a. The increase in the relative intensity of the band with a maximum at 800 cm^{-1} compared to the intensity of the band at 890 cm^{-1} in the spectrum of the LAS glass-ceramic suggests that the rate of liquid–liquid phase separation in this glass during this heat treatment is higher than that in the LAS_{ox} glass, i.e., neutral conditions of glass melting speed up the development of liquid–liquid phase separation in this glass. Crystallization of the titanate phases of anatase and tieilite are sped up in the LAS glass ceramized by the heat treatment at the second stage at 780 °C as compared with the LAS_{ox} glass, see Figure 10b. The same tendency remains true for glass-ceramics obtained by heat treatment with the second stage at 900 °C, see Figure 10c. The higher intensity of the anatase peaks in the Raman spectrum of the LAS_{ox} glass-ceramic obtained by heat treatment at 950 °C, see Figure 10d, means that anatase crystallization in the LAS_{ox} glass reaches its maximum and will decrease at higher temperatures while the maximum anatase content in the LAS glass-ceramics was reached at a previous holding temperature. The traces of anatase crystals remain in the LAS_{ox} glass-ceramic obtained by heat treatment at 1000 °C while the LAS glass-ceramic does not show spectral features of this phase, and intensities of peaks assigned to tieilite crystals are higher for the LAS glass-ceramic than for the LAS_{ox} one, see Figure 10e. Figure 10f shows that the LAS glass-ceramic prepared by heat treatment at 1200 °C demonstrates rutile bands of higher intensities as compared with the LAS_{ox} glass-ceramic.

Though Raman spectra of precursor glasses and glasses heat-treated at 680 °C for 6 h are similar to each other, a comparison of the Raman spectra of the materials obtained by two-stage heat treatments unambiguously indicates that, despite the fact that the sequence of transformations in the titania-containing phase is independent of the redox conditions of glass melting, the rate of these transformations is significantly higher at ceramming of the LAS glass.

3.4. Morphology Characterization by SEM

SEM analysis (Figure 11a) reveals the presence of inhomogeneous regions within the bulk of the initial amorphous LAS glass. The calculated size distribution shows a broad profile, with the mean size of the inhomogeneous regions being approximately 27 nm (Figure 12). These inhomogeneities may indicate liquid–liquid phase separation during glass formation. The broad size distribution may arise from overlapping size distributions of chemically distinct regions. Such regions are likely precursors to the crystallization of various phases upon subsequent heat treatments.

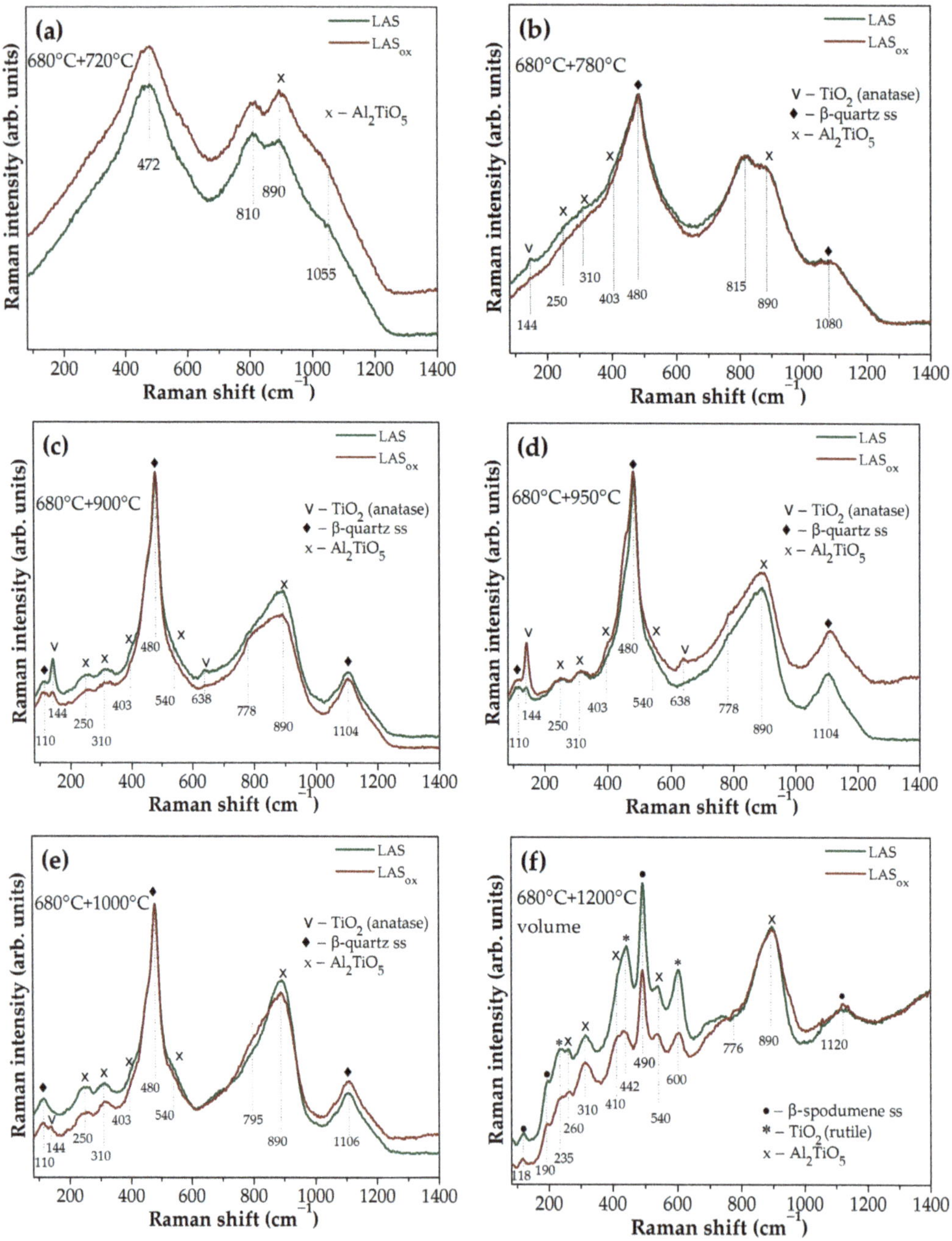

Figure 10. Raman spectra the LAS and the LAS$_{ox}$ glass-ceramics obtained by two-stage heat treatments at: (**a**) 680 °C + 720 °C; (**b**) 680 °C + 780 °C; (**c**) 680 °C + 900 °C; (**d**) 680 °C + 950 °C; (**e**) 680 °C + 1000 °C; (**f**) 680 °C + 1200 °C. The timeof heat treatment at each stage is 6 h. The spectra are shifted for the convenience of observation.

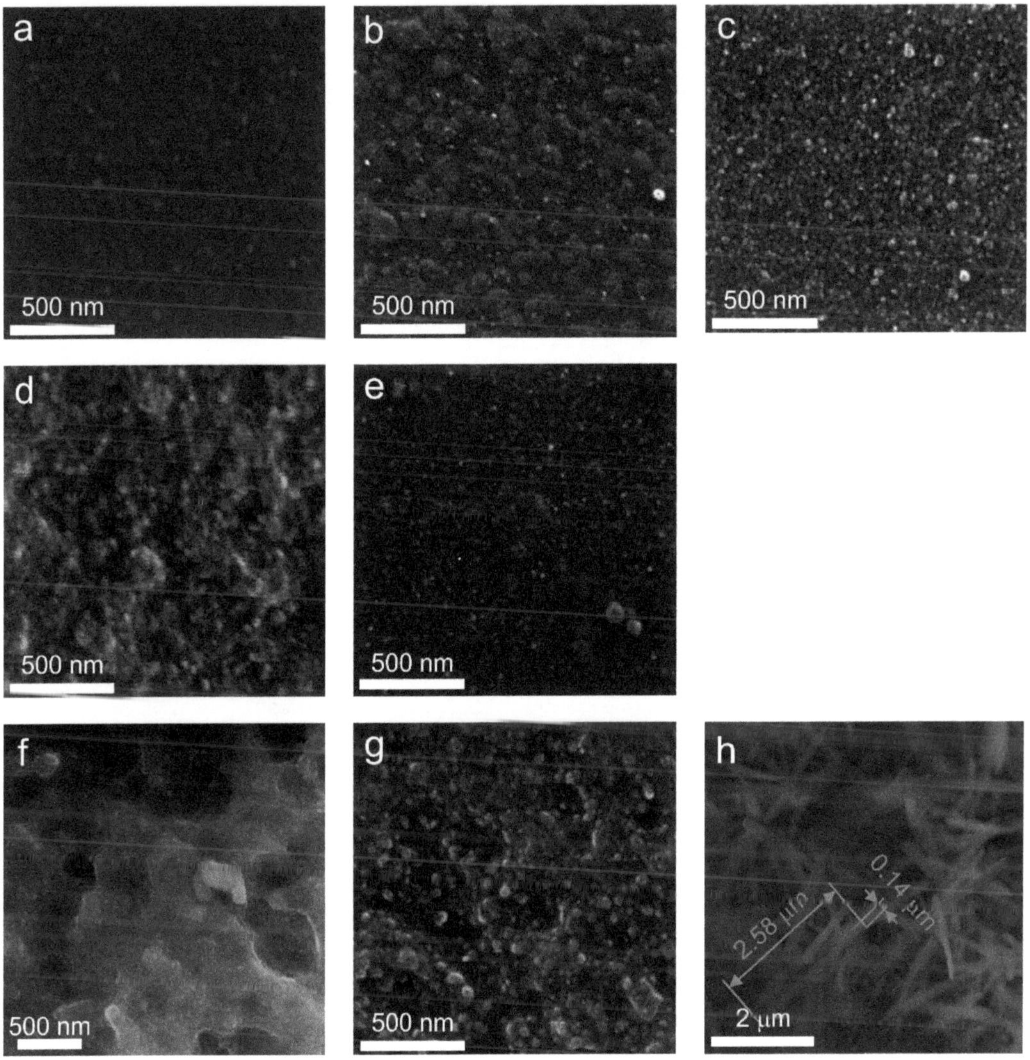

Figure 11. SEM images of the samples of: (**a**) the initial LAS glass; (**b**) the LAS glass after heat treatment at 680 °C; (**c**) the LAS glass ceramized at 680 °C and at 750 °C, (**d**) the LAS glass ceramized at 680 °C and at 780 °C; (**e**) the LAS_{ox} glass ceramized at 680 °C and at 780 °C; (**f**) the LAS glass ceramized at 680 °C and at 800 °C; (**g**) the LAS glass ceramized at 680 °C and at 1000 °C; (**h**) the LAS glass ceramized at 680 °C and at 1200 °C. The holding time is 6 h.

Following the single-step heat treatment at 680 °C for 6 h, the LAS glass remains X-ray amorphous. However, its SEM image (Figure 11b) shows increased inhomogeneity, clearly distinguishing two types of regions. The first type consists of small, spherical, bright regions with a narrow size distribution averaging 16 nm (Figure 12). The second predomonant type features larger, irregular, dark regions with an average size of 74 nm (Figure 12).

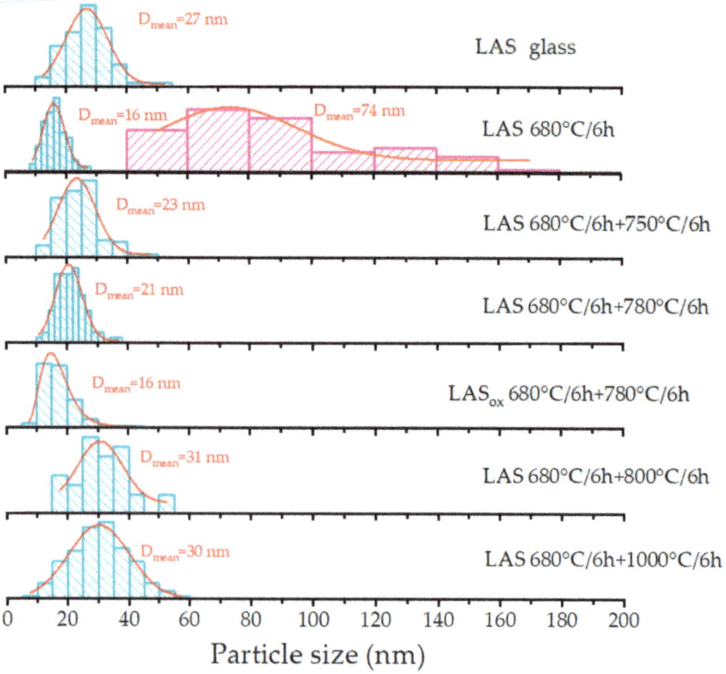

Figure 12. Particle size distributions in the initial LAS glass and LAS and LAS_{ox} glass-ceramics calculated from SEM data presented in Figure 11. Heat-treatment schedules and mean crystallite sizes are listed in figure.

After two-stage heat treatment at 680 °C for 6 h and 750 °C for 6 h, the large inhomogeneous regions are not seen on the SEM image anymore. Numerous spherical particles appear in the SEM image of the LAS glass-ceramic, averaging 23 nm in size (Figure 11c). The broad shape of the size distribution suggests overlap from multiple chemically distinct particle populations, see Figure 12. XRD analysis confirms the crystallization of the spinel phase with a mean crystallite size of 4.5 nm.

The heat treatment at 680 °C for 6 h and 780 °C for 6 h promotes significant crystallization of the LAS glass. The SEM image of this glass-ceramic is shown in Figure 11d. XRD analysis confirms the formation of three crystalline phases, β-quartz ss, spinel, and Al_2TiO_5, see Figure 7c, with mean crystallite sizes of 8 nm, 26 nm and ~6 nm, respectively, see Tables 2 and 3. The particle size distribution suggests the possible overlap of multiple size distributions corresponding to different phases, see Figure 12. The mean particle size is approximately 21 nm, which is close to the mean size of the predominant crystalline phase of β-quartz ss.

The two-step heat treatment at 680 °C and 800 °C for 6 h results in more extensive crystallization of the LAS glass, see Figure 7e. XRD analysis confirms the presence of crystalline phases of β-quartz ss, spinel, and Al_2TiO_5. The SEM image reveals significant etching of the material, characterized by numerous large caverns, which may indicate silica depletion, see Figure 11f. The mean particle size is 31 nm, see Figure 12, which is also close to the mean size of the predominant crystalline phase of β-quartz ss, which is 25 nm, see Table 2.

Following the two-step heat treatment at 680 °C and 1000 °C for 6 h, XRD analysis shows the formation of β-quartz ss, mullite, and Al_2TiO_5. Signs of the residual glass phase become less prominent. The particle size distribution, similar to previous samples, exhibits a broad profile, likely due to overlapping distributions from different phases, with an average particle size of 30 nm. The size of the predominant crystalline phase of β-quartz ss is 28 nm, see Table 2.

Finally, after two-step heat treatment at 680 °C and 1200 °C for 6 h, the SEM image of the opaque LAS sample shows the presence of micron-sized, needle-like crystals and agglomerates of spherical crystals, see Figure 11h. The needle-like crystals are very similar to crystals of β-spodumene ss that were crystallized in the lithium aluminosilicate glass of a different composition during its heat treatment at 1350 °C [12]. XRD analysis confirms the crystallization of β-spodumene (ss), mullite, and Al_2TiO_5 phases, see Figure 7e.

The influence of redox conditions of glass melting on the phase assamblage of glass-ceramics is revealed by the comparison of SEM images of LAS and LAS_{ox} glass-ceramics obtained by two-stage heat treatment at 680 °C and 780 °C for 6 h. In accordance with XRD data, see Figure 8a, the SEM image of the LAS_{ox} glass-ceramic shows a significantly reduced crystallinity fraction, see Figure 11e. The particle size distribution for this sample also indicates overlapping distributions, with a mean particle size of 16 nm, which is smaller than the mean particle size of the LAS glass-ceramic. Considering that the main contribution to the increase in mean crystallite size comes from larger crystals of β-quartz ss, the smaller mean crystallite size in the LAS_{ox} glass-ceramic indicates a smaller fraction of these crystals.

3.5. Optical Spectroscopy

Light losses in phase separated glasses and glass-ceramics are determined by absorption due to coloring ions and light scattering on the interfaces of regions of inhomogeneity inherent to such materials. It is generally accepted that the coloration of glass-ceramics containing titania as a nucleating agent is mainly caused by intervalent charge transfer transitions between titanium and iron impurity ions in different oxidation states [8,53]. Though the influence of absorption due to the Fe^{3+}-Ti^{3+} and Fe^{2+}-Ti^{4+} intervalence charge transitions on absorption spectra of our materials cannot be excluded, only absorption due to titanium ions was considered in this study. A similar approach was suggested in [53]. Titanium ions in glasses exist in two oxidation states. The presence of Ti^{3+} and Ti^{4+} ions, and Ti^{3+}-Ti^{4+} pairs defines the absorption of glasses under study. Ti^{4+} ion has an electronic configuration of $3d^0$ and does not demonstrate absorption bands due to d–d transitions in the ligand field. Ti^{4+} ions participate in O-Ti^{4+} charge transfer bands positioning in the UV range of spectrum at ~300 nm [54] and in intervalence charge transfer Ti^{4+}-Ti^{3+} transitions responsible for coloration in the visible spectral range [55]. Ti^{3+} ions have an electronic configuration of $3d^1$ and demonstrate one broad band in the visible range of spectrum due to the $^2T_{2g} \rightarrow E_g$ transition of the Ti^{3+} ions in octahedral symmetry. The band often has a shoulder at the long-wavelength side of the absorption band caused by the Jahn–Teller effect [23,25–27,55]. The absorption band of Ti^{3+} ion in tetrahedral (T_d) symmetry caused by the $^2T_{2g} \rightarrow E_g$ transition is located in the near-infrared region of the spectra [55]. The O-Ti^{3+} charge transfer band is expected in UV range of spectrum at ~240 nm [54].

Absorption spectra of initial and heat-treated glasses are shown in Figure 13a–h. For the convenience of comparison, the spectra of the LAS and LAS_{ox} glasses and glass-ceramics obtained by the same heat-treatment schedule are presented in Figure 14a–h. The spectra are formed by the absorption edge in the UV spectral range, see Figure 13e,f, intense absorption of OH^- groups spanning from 2700 nm to 3300 nm, see Figure 13g,h, and light losses in the visible and near-IR spectral range, see Figure 13a–c, which have different origin in the LAS and the LAS_{ox} glass-ceramics. Note that the LAS_{ox} glass-ceramics obtained by heat treatments at 950 °C and 1000 °C at the second stage cracked during heat treatment, and their spectra were not recorded.

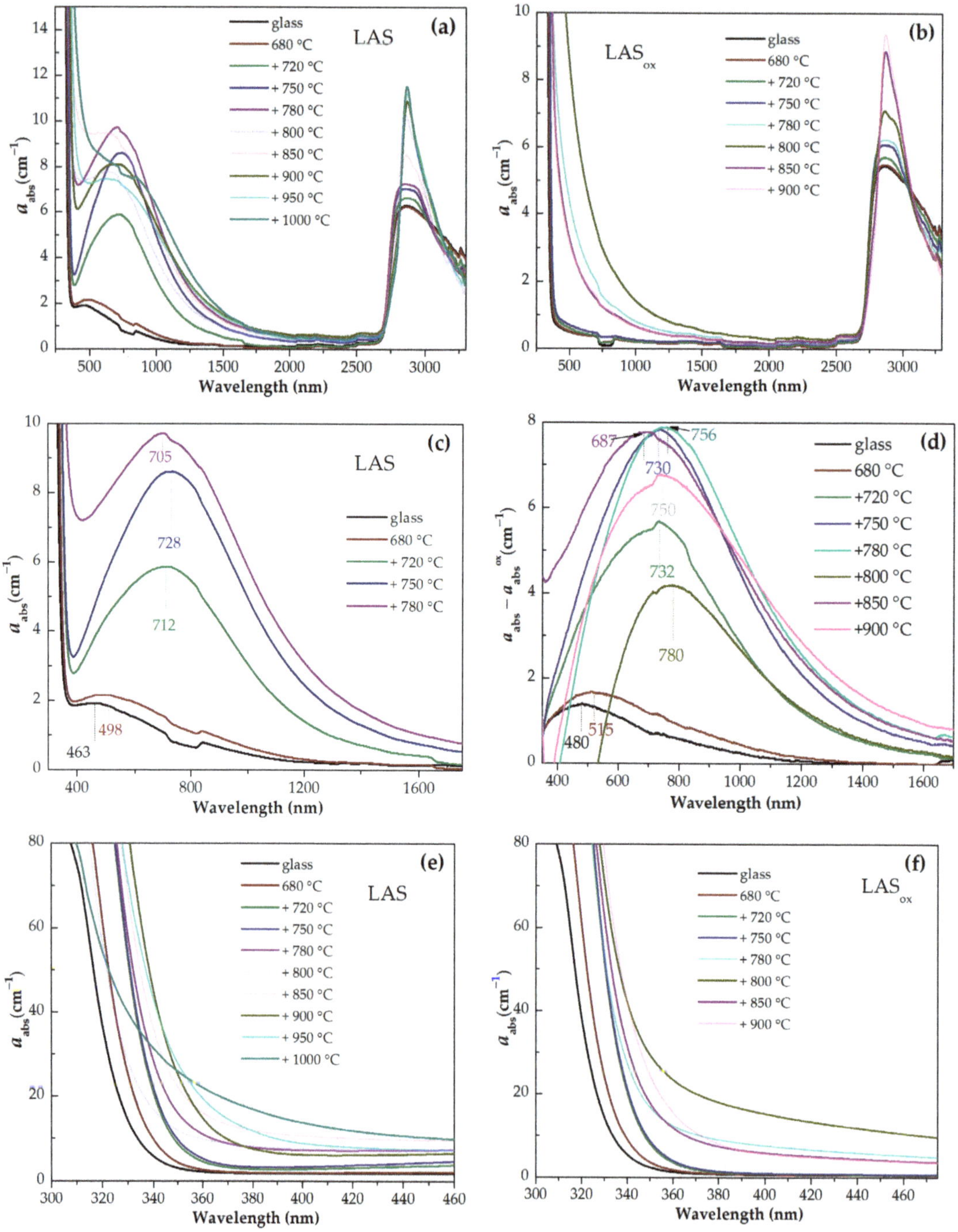

Figure 13. *Cont.*

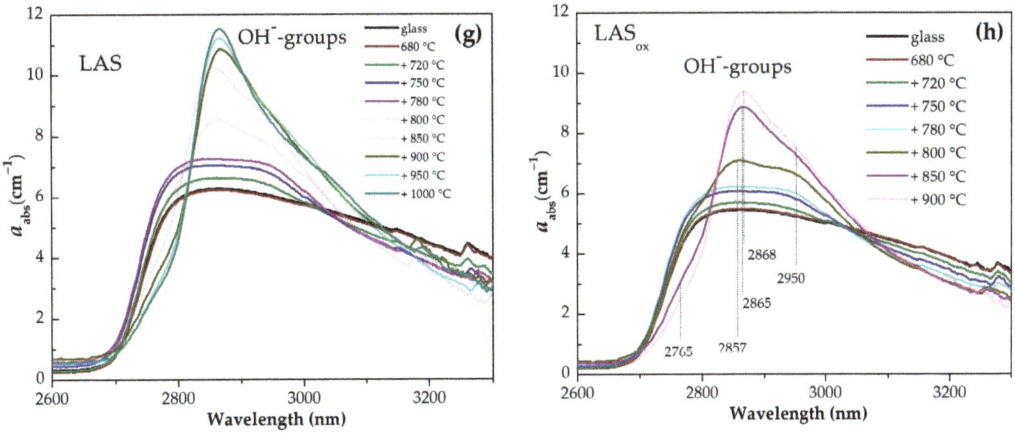

Figure 13. Absorption spectra in different spectral ranges of the samples of: (**a**) the initial LAS glass and glass heat-treated up to 1000 °C in the spectral range from 250 nm to 3300 nm; (**b**) the initial LAS_{ox} glass and glass heat-treated up to 900 °C in the spectral range from 250 nm to 3300 nm; (**c**) the LAS glass and glass heat-treated up to 780 °C in the spectral range from 300 nm to 1750 nm; (**d**) the subtracted spectra; (**e**) the LAS glass and glass heat-treated up to 1000 °C in the spectral range from 300 nm to 460 nm; (**f**) the LAS_{ox} glass and glass heat-treated up to 900 °C in the spectral range from 300 nm to 475 nm; (**g**) the LAS glass and glass heat-treated up to 1000 °C in the spectral range from 2600 nm to 3300 nm; (**h**) the LAS_{ox} glass and glass heat-treated up to 900 °C in the spectral range from 2600 nm to 3300 nm. The duration of each hold is 6 h.

For the initial LAS glass, the UV absorption edge is observed at ~330 nm. In the spectrum of the glass heat-treated at the nucleation stage at 680 °C for 6 h it moves to longer wavelengths by 5 nm. In the spectrum of glass-ceramics obtained by the two-stage heat treatment with a temperature of 720 °C at the second stage, the position of the absorption edge is found at ~343 nm and it is nearly the same after heat treatments at 750 °C and 780 °C. After heat treatment with a temperature of 800 °C at the second stage, the absorption edge shifts to shorter wavelengths, to 334 nm. After increasing the temperature of the second stage to 850 °C and 900 °C, the absorption edge shifts again to longer wavelengths. Its position is equal to 347 nm and 351 nm, respectively. After heat treatments at 950 °C and 1000 °C, the position of the absorption edge shifts again to shorter wavelengths, see Figure 13e.

The position of the absorption edge for the LAS_{ox} glass and its variation with the heat-treatment temperature are surprisingly similar to those of the LAS glass, see Figure 13e,f and Figure 14a–e,g,h. The position of the absorption edge differs only for glass-ceramics obtained by the heat treatment at the second stage at 800 °C, see Figure 14f. The absorption edge in the spectrum of this LAS_{ox} glass-ceramic is shifted to longer wavelengths compared to its position in the spectrum of the LAS sample. The nonmonotonic transmittance variation during the crystallization of glasses inclined to liquid–liquid phase separation was described in ref. [56] and assigned to incoherent scattering that takes place in a material containing amorphous and crystallized regions of inhomogeneity. The authors of ref. [56] demonstrated that the extinction coefficient can reach a maximum when the crystallinity fraction is 0.5–1.0 and then decrease due to the presence of elements of ordering in the relative position of the crystallites.

The same raw materials were used for the preparation of both glasses, which means that the iron content in both glasses was similar. The absorption edge in precursor glass and in glass-ceramics originates from $O^{2-} \rightarrow Ti^{4+}$ and $O^{2-} \rightarrow Ti^{3+}$ charge transfer bands [54]. Taking into account that the $O^{2-} \rightarrow Ti^{4+}$ charge transfer band is located at longer wavelengths than the $O^{2-}\rightarrow Ti^{3+}$ one, it can be suggested that the content of Ti^{4+} ions in the LAS and the LAS_{ox} glasses is very similar, i.e., the content of Ti^{3+} in the LAS glass is rather low. In glass-ceramics containing β-quarts ss, scattering losses are superimposed with the absorption edge.

A broad absorption band is found in the spectrum of the LAS precursor glass. Its intensity increases after heat treatment at the nucleation stage and with the increasing of the heat-treatment temperature at the second stage up to 780 °C, see Figure 13a,c. After further heat treatments the absorption band becomes broader, and its intensity somewhat decreases. The appearance of this absorption band can be explained by Ti^{3+} ions distributed between different amorphous and crystalline phases. Since the absorption edge and light scattering in the obtained multiphase materials are superimposed on the short-wave part of the absorption spectrum, absorption due to these losses was subtracted from the experimental absorption spectra of the LAS samples using the corresponding absorption spectra of LAS_{ox} samples for substraction. The difference spectra are shown in Figure 13d. The spectrum of the precursor glass is a typical absorption spectrum of Ti^{3+} ions in silicate glasses with a broad peak with a maximum at ~480 nm attributed to d–d transition $^2T_{2g}\rightarrow{}^2E_g$ of $3d^1$ electron of Ti^{3+} ions in octahedral coordination in the ligand field affected by the Jahn–Teller effect [25–27]. The tail with a maximum at ~800 nm is assigned to intervalence charge transfer in Ti^{3+}-Ti^{4+} pairs [57,58]. The absorption intensity increases after the preliminary heat treatment, the band maximum shifts to ~500 nm, which is connected with the participation of Ti^{3+} ions in the liquid–liquid phase separation and entering into the aluminotitanate amorphous regions. With the increase in the heat-treatment temperature at the second stage, the broadband absorption in the visible and near-IR region intensifies, and its maximum shifts to ~730 nm in the spectra of glass-ceramics obtained by heat treatments at 750 °C and 780 °C at the second stage. Similar to the absorption of Ti^{3+} ions in corrund crystals [59], this absorption is attributed to Ti^{3+} ions in the ligand field of octahedral symmetry (the shoulder at ~480 nm) and Ti^{3+}-Ti^{4+} pairs in crystals of γ-Al_2O_3. In the spectrum of corrundum, the Ti^{4+} ions in Al^{3+} sites are considered to be charge-compensated by Al vacancies with one vacancy for every three Ti^{4+} ions [60]. An increase of the lattice parameters of γ-Al_2O_3 with heat-treatment temperature found by the XRD analysis can be due to the entering of the crystals of γ-Al_2O_3 by Ti^{3+} ions (ionic radius in octahedral coordination is 0.81 Å) and Ti^{4+} ions (ionic radius in octahedral coordination is 0.745 Å). Note that the ionic radii of Al^{3+} in octahedral coordination is 0.675 Å.

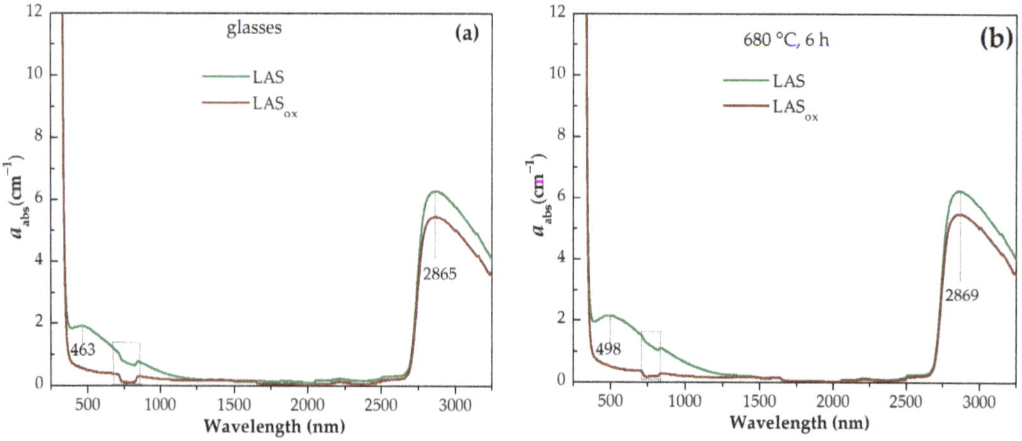

Figure 14. *Cont.*

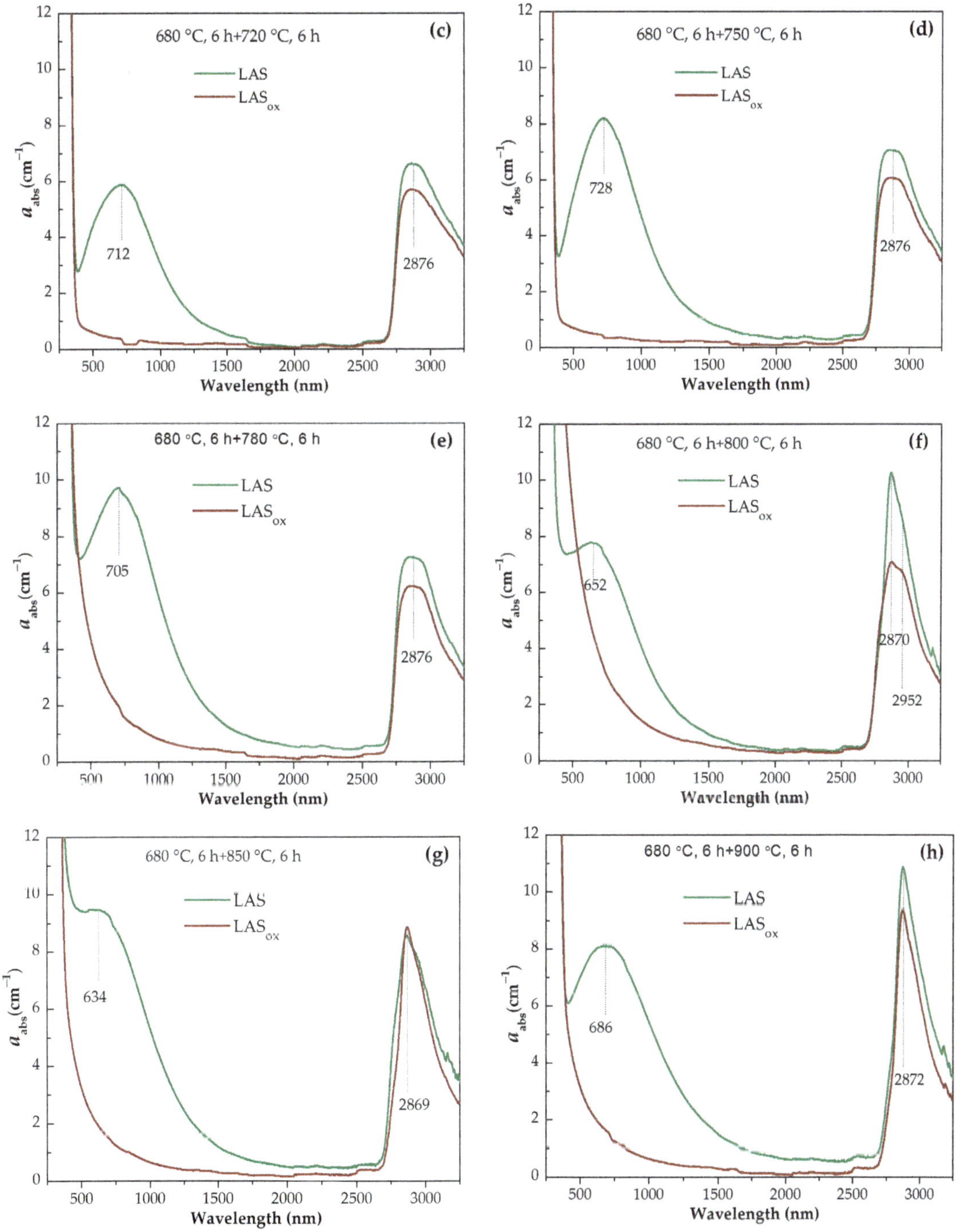

Figure 14. Absorption spectra of the samples of: (**a**) the initial LAS and LAS_{ox} glasses; (**b**) the LAS and LAS_{ox} glasses heat-treated at 680 °C; (**c**) the LAS and LAS_{ox} glasses heat-treated at 680 °C + 720 °C; (**d**) the LAS and LAS_{ox} glasses heat-treated at 680 °C + 750 °C; (**e**) the LAS and LAS_{ox} glasses heat-treated at 680 °C + 780 °C; (**f**) the LAS and LAS_{ox} glasses heat-treated at 680 °C + 800 °C; (**g**) the LAS and LAS_{ox} glasses heat-treated at 680 °C + 850 °C; (**h**) the LAS and LAS_{ox} glasses heat-treated at 680 °C + 900 °C. The duration of each hold is 6 h.

The absorption band shape changes with the raising of the heat-treatment temperature. The spectrum is still mainly formed by the absorption of Ti^{3+}-Ti^{4+} pairs. Taking into account the crystallization of β-quartz ss and tieilite, Al_2TiO_5, this change in the shape of the absorption band can be connected with the distribution of titanium ions between the crystalline phases. Spinel does not crystallize in glass-ceramics obtained by heat treatments at 850–1000 °C at the second stage. So the broadband spectra of glass-ceramics obtained by these heat treatments are connected with titanium ions in β-quartz ss and tieilite.

The light losses in the LAS_{ox} glass-ceramics are mostly caused by light scattering, which has a non-monotoneous dependence on the heat-treatment temperature, as was mentioned above.

Figure 13g,h show wide asymmetric absorption bands in the spectral region from ~2700 nm to 3300 nm caused by the presence of OH-groups in glasses and glass-ceramics. Heat treatment at the nucleation stage at 680 °C for 6 h has no effect on the shape and intensity of this band. Two-stage heat treatments with a temperature from 720 °C to 780 °C at the second stage cause a successive change in the spectrum of OH-groups. The intensity in the range from 2750 nm to 3050 nm increases and the intensity in the range from 3050 nm to 3300 nm decreases. Further narrowing of the OH-groups absorption band and growth of its intensity is observed in spectra of glass-ceramics obtained during heat treatments in the range of crystallization temperatures from 800 °C to 1000 °C. Comparison of the absorption bands of OH-groups in glasses melted in different redox conditions and in corresponding glass-ceramics showed that the intensities of the absorption bands of OH-groups are higher in inital and heat-treated LAS glass than in LAS_{ox} samples. The absorption band shape is the same for the materials obtained by the same heat-treatment shedule, see Figure 14a–h, which is explained by the similarity of their phase compositions.

3.6. Estimation of the Coefficient of Thermal Expansion

Table 4 lists the coefficients of thermal expansion (CTE) for the initial and heat-treated glasses of the LAS and the LAS_{ox} compositions in the temperature range between 20 °C and 320 °C. The CTE values of precursor glasses are similar and equal to ~4.3×10^{-6} K^{-1}. After the nucleation heat treatment at 680 °C for 6 h the CTE values slightly decrease to become ~4.1×10^{-6} K^{-1}. There is a gradual increase of CTE values with the increase in the ceramming temperature to 720 °C and then to 750 °C. Glass-ceramics based on spinel nanocrystals and prepared by the consequent three-stage heat treatment schedule 680 °C, 6 h + 720 °C, 6 h + 750 °C, 6 h have the highest CTE value of ~5.2×10^{-6} K^{-1}. The CTE value starts to decrease as the first portions of β-quartz ss crystallize (heat treatment with the last hold at 780 °C for 6 h) and reaches the lowest value of ~0.3×10^{-6} K^{-1} for the glass-ceramics obtained by ceramming at the last stage of heat treatment at 950 °C for 6 h. After heat treatment of the precursor glass at 1200 °C, the CTE increases again to a value of 1.46×10^{-6} K^{-1} due to the crystallization of β-spodumene ss. The dependences of the change in the CTE value on the temperature of heat treatment for samples obtained by the crystallization of glasses melted under neutral and oxidizing conditions are similar, which can be explained by the similarity of their phase assemblage.

Table 4. Coefficients of thermal expansion (CTE) for precursor and heat-treated glasses of the LAS and the LAS_{ox} compositions in the temperature range from room temperature to 320 °C.

Heat-Treatment Regime	**$CTE_{20\text{-}320}$ ($\times 10^{-6}$ K^{-1})**	
	LAS	**LAS_{ox}**
Precursor glasses	4.35 ± 0.08	4.25 ± 0.08
680 °C, 6 h	4.19 ± 0.08	4.21 ± 0.08
680 °C, 6 h + 720 °C, 6 h	4.82 ± 0.09	4.57 ± 0.09

Table 4. *Cont.*

Heat-Treatment Regime	$CTE_{20\text{-}320}$ ($\times 10^{-6}$ K^{-1})	
	LAS	LAS_{ox}
680 °C, 6 h + 720 °C, 6 h + 750 °C, 6 h	5.36 ± 0.10	5.18 ± 0.10
680 °C, 6 h + 720 °C, 6 h + 750 °C, 6 h + 780 °C, 6 h	4.30 ± 0.08	4.85 ± 0.09
680 °C, 6 h + 780 °C, 6 h	-	4.85 ± 0.10
720 °C, 6 h + 800 °C, 6 h	1.15 ± 0.02	1.24 ± 0.01
720 °C, 6 h + 850 °C, 6 h	0.61 ± 0.01	0.45 ± 0.08
680 °C, 6 h +720 °C, 6 h +750 °C, 6 h +950 °C, 6 h	0.30 ± 0.01	0.19 ± 0.01
680 °C, 6 h +1200 °C, 6 h	1.49 ± 0.03	1.42 ± 0.03

4. Discussion

Glasses melted in neutral and oxidizing conditions are X-ray amorphous and inhomogeneous according to SEM microscopy findings. Inhomogeneity regions have a broad size distribution with a mean size of 27 nm. In accordance with previous studies [19,20], it is suggested that liquid–liquid phase separation occurs during glass melt casting and annealing. Absorption spectra of glasses melted in different redox conditions are different by the appearance of a broadband absorption in the LAS glass mainly assigned to Ti^{3+} ions in distorted octahedral coordination and Ti^{3+}-Ti^{4+} pairs in the glass structure. One may suggest that these ions are distributed between the lithium aluminosilicate glass matrix and amorphous regions of inhomogeneity composed of octahedrally coordinated titanium and aluminum ions even with the formation of Ti^{3+}-Ti^{4+} clusters [18]. Based on the similarity of XRD patterns, Raman spectra and glass transition temperatures, it was speculated that glasses melted in neutral and oxidizing conditions have a similar structure because of the low content of Ti^{3+} ions. Nevertheless, the role of Ti^{3+} ions in liquid–liquid phase separation and crystallization of LAS glass is clearly seen.

The glasses obtained by heat treatment at 680 °C for 6 h do not contain any crystalline phase, their XRD patterns are similar, and their Raman spectra are near similar to each other and to those of precursor glasses. However, there is a tremendous difference in the behavior of their DSC curves, their morphology revealed in the SEM study and their absorption spectra as compared with those of precursor glasses. It allows us to conclude that liquid–liquid phase separation continues to develop in both glasses during the nucleation heat treatment.

According to data of SEM analysis, two types of amorphous regions with mean sizes of 16 and 74 nm are formed in the glass during heat treatment at 680 °C for 6 h. A similar structure formation was not found in previously studied glasses of the lithium aluminosilicate system. The formation of large amorphous aluminotitanate regions of inhomogeneity in magnesium ([20] and refs therein) and zinc ([20] and refs therein) aluminosilicate glasses nucleated by titania was previously revealed by a combination of small-angle X-ray scattering and Raman spectroscopy data. Those materials also contained smaller size phase-separated regions enriched in aluminates of corresponding cations ([20] and refs therein). One can suggest a similar character of phase separation in glasses under study and suppose that Ti^{3+} ions are distributed between large aluminotitanate amorphous regions, smaller size aluminate regions and residual glass.

Judging by the data of Raman spectroscopy, glasses cerammed using a two-stage heat treatment with a temperature at the second hold of 720 °C have a well-developed liquid–liquid phase separated structure. During this phase separation, spinel nanocrystals appear in both glasses, and the composition of the residual lithium aluminosilicate glass slightly changes, which is manifested by a change in the position of the amorphous halo in the XRD pattern and the position of the band at ~480 cm^{-1} in the Raman spectrum. Spinel is the sole crystalline phase in transparent glass-ceramics obtained by heat treatments at the second stage at 720 °C and 750 °C. Upon crystallization of spinel, titanium ions Ti^{3+} enter its structure, which is clearly seen in the absorption spectra of the LAS glass-ceramics. The change of the position of the maximum of absorption band to longer wavelengths suggests that absorption spectrum in glass-ceramics is mainly formed by the Ti^{3+}-Ti^{4+} intervalent charge transfer band. This is a spectroscopic confirmation of the presence of Ti^{4+} ions in the spinel structure.

Therefore, based on the DSC, SEM and XRD data, it was concluded that preliminary heat treatment at 680 °C, resulting in the formation of a reach phase-separated structure, provokes spinel crystallization during further heat treatments between 720 °C and 800 °C. According to the similarity of absorption spectra of these glass-ceramics and spectra of Ti^{3+}-doped Al_2O_3, it was speculated that Ti^{3+} ions enter into spinel nanocrystals. Taking into consideration that, according to the findings of the Raman spectroscopy, the rate of phase transformations in the titania-containing phase in glass melted in neutral conditions is significantly higher than in glass melted in oxidizing conditions, it was suggested that Ti^{3+} ions become a component of amorphous aluminotitanate phase-separated regions.

Previously, the decrease in intensity of the first broad peak in the DSC curves of glasses nucleated at 700 °C and its near disappearance from DCS curves of glasses preliminarily heat-treated at 720 °C for 6 h was mentioned. The sample preliminarily heat-treated at 720 °C already contained spinel nanocrystals. Therefore, in this sample only high temperature phases of β-quartz ss, tieilite and β-spodumene ss crystallized, which resulted in the appearance of the second and the third peaks and the absence of the first peak on the DSC curve.

A specific feature of the morphology of the materials under study is the similarity of the mean size of inhomogeneous regions in the precursor glass and in glass-ceramics obtained by two-stage heat treatments at temperatures up to 1000 °C. Sizes of β-quartz ss crystallized in both glasses in the temperature range from 780 °C to 1000 °C are similar to each other. Our findings are in line with the results presented in ref. [38] and devoted to the study of the lithium aluminosilicate glass nucleated by a mixture of TiO_2 and ZrO_2, where the constancy of the size of β-quartz ss crystallized in the temperature range from 860 °C to 960 °C was revealed and discussed.

The comparison of XRD patterns of the LAS and the LAS_{ox} glass-ceramics obtained at 1200 °C demonstrates that LAS glass-ceramic contains a larger fraction of rutile and a smaller fraction of mullite than the LAS_{ox} glass-ceramic. Previously it was revealed the difference in lattice parameters of rutile crystallized in zinc aluminosilicate glasses nucleated by TiO_2 and melted in different redox conditions [34]. In ref. [34] this difference was explained by the presence of Ti^{3+} ions in the rutile structure. Here, it is suggested that in lithium aluminosilicate glass-ceramics Ti^{3+} ions also participate in the formation of rutile nanocrystals, which results in the facilitation of their crystallization. Note that the role of small additives of transition metal ions in kinetics of crystallization in glass-ceramics was discussed in ref. [18].

The rate of phase transformations resulted in the crystallization of β-quartz ss is significantly higher during the heat treatment of the glass melted in neutral conditions. However, the sequence of phase transformations and phase assemblage of glass-ceramics melted in different redox conditions are independent of the redox conditions of glass melting. This is very promising for the development of transparent lithium aluminosilicate glass-ceramics with close-to-zero thermal expansion coefficient and doped with various functional ions in the lower oxidation states. Crystallization of β-quartz ss leads to a decreasing of the value of the thermal expansion coefficient. The lowest coefficient of thermal expansion was obtained by the heat treatment at 950 °C. It was $\sim 0.3 \times 10^{-6}$ K^{-1}.

5. Conclusions

The glass of the lithium aluminosilicate system containing titanium oxide as a nucleating agent was melted with and without the addition of As_2O_3, i.e., under oxidizing and neutral conditions. The glasses were heat-treated in the temperature range from 680 °C to 1300 °C to obtain glass-ceramics. Glass-ceramics developed from glasses melted without the addition of As_2O_3 are black-colored, and those fabricated from glasses melted with addition of As_2O_3 are colorless.

Transparent glass-ceramics based on β-quartz ss and/or γ-Al_2O_3 nanocrystals and opaque glass-ceramics based on β-spodumene ss nanocrystals were obtained. Crystallization of β-quartz ss was accompanied by nanocrystals of tieilite, Al_2TiO_5. Crystallization of β-spodumene ss was accompanied by tieilite, mullite (at 1100 °C) and rutile (at 1200 °C).

Heat treatment at the nucleation stage causes crystallization of γ-Al_2O_3 with a spinel structure and an increase in the crystallization temperature of β-quartz ss and Al_2TiO_5 and decrease in the temperature of β-spodumene ss crystallization. Glasses melted under neutral conditions lose transparency at higher secondary heat-treatment temperatures than glasses melted under oxidizing conditions.

X-ray amorphous precursor glasses have an inhomogeneous structure with a mean size of regions of inhomogeneity of ca. 27 nm. Preliminary heat treatment at 680 °C results in the formation of a reach bimodal liquid–liquid phase separated structure with mean sizes of inhomogeneous regions of ca. 16 nm and 74 nm. Titanium ions enter both types of inhomogeneous regions. An appearance of a bimodal structure of inhomogeneous regions in the result of heat treatment at the nucleation stage suggests the development of a three-phase immiscibility in the lithium aluminosilicate glass with a high excess of alumina over lithium oxide, i.e., the formation of two amorphous phases, the aluminotitanate phase and the aluminate phase, dispersed in a high-silica matrix. It is the first time that three-phase immiscibility is revealed in glasses of the lithium aluminosilicate system.

It is the first time that the formation of nanocrystals of γ-Al_2O_3 with a spinel structure appeared during heat treatments at the second stage in the temperature interval between 720 °C and 800 °C was documented. Titanium ions are located in the spinel structure as Ti^{3+} ions in the ligand field of octahedral symmetry and in Ti^{3+}-Ti^{4+} pairs. Ti^{3+} ions also enter amorphous aluminotitanate phase-separated regions and facilitate phase transformations in this phase with formation of tieilite, anatase and rutile.

The phase assemblage and sequence of phase transformations is independent of the redox conditions of glass melting, however, the rate of these transformations is significantly higher when ceramming the glass melted in neutral conditions. This is very promising for the development of transparent glass-ceramics with close-to-zero thermal expansion coefficient and doped with various functional ions in lower oxidation states.

The variation of the coefficient of thermal expansion with heat-treatment temperature reflects the variation of the phase composition of the developed materials, does not depend on the redox conditions of glass melting and reaches as low value as ~0.3×10^{-6} K^{-1} for glass-ceramics based on nanocrystals of β-quartz ss.

Author Contributions: Conceptualization, O.D. and A.Z.; methodology, I.A., K.B., M.T. and A.B.; software, A.B.; validation, I.A. and O.D.; formal analysis, I.A., A.B. and A.V.; investigation, S.M., S.Z., M.T., A.V., I.A. and O.D.; resources, K.B.; data curation, I.A., A.B. and A.V.; writing—original draft preparation, O.D., I.A. and A.B.; writing—review and editing, O.D., A.Z. and G.S.; visualization, S.M., A.V. and A.B.; supervision, O.D.; project administration, O.D.; funding acquisition, O.D. and G.S. All authors have read and agreed to the published version of the manuscript.

Funding: This research received no funding.

Data Availability Statement: The original contributions presented in this study are included in the article. Further inquiries can be directed to the corresponding author.

Conflicts of Interest: The authors declare no conflicts of interest.

References

1. Höland, W.; Beall, G.H. *Glass-Ceramic Technology*; Wiley: Hoboken, NJ, USA, 2012; pp. 92–97, 269–279.
2. Bach, H.; Krause, D. *Low Thermal Expansion Glass Ceramics*; Springer-Verlag: Berlin/Heidelberg, Germany, 2005.
3. Petzoldt, J.; Pannhorst, W. Chemistry and structure of glass-ceramic materials for high precision optical applications. *J. Non-Cryst. Solids* **1991**, *129*, 191–198. [CrossRef]
4. Stookey, S.D. Method of Making Ceramics and Product Thereof. U.S. Patent 2,920,971, 12 January 1960.
5. Beall, G.H.; Karstetter, B.R.; Rittler, H.L. Crystallization and chemical strengthening of stuffed β-quartz glass-ceramics. *J. Am. Ceram. Soc.* **1967**, *5*, 181–190. [CrossRef]
6. Zanotto, E.D. A bright future for glass-ceramics. *Am. Ceram. Soc. Bull.* **2010**, *89*, 19–27.
7. Fernandez-Martin, C.; Bruno, G.; Crochet, A.; Ovono, D.O.; Comte, M.; Hennet, L. Nucleation and growth of nanocrystals in glass-ceramics: An in situ SANS perspective. *J. Am. Ceram. Soc.* **2012**, *95*, 1304–1312. [CrossRef]
8. Nakane, S.; Kawamoto, K. Coloration mechanism of Fe ions in β-quartz *s.s.* glass-ceramics with TiO_2 and ZrO_2 as nucleation agents. *Front. Mater.* **2017**, *4*, 14–19. [CrossRef]
9. Fu, Q.; Wheaton, B.R.; Geisinger, K.L.; Credle, A.J.; Wang, J. Crystallization, microstructure, and viscosity evolutions in lithium aluminosilicate glass-ceramics. *Front. Mater.* **2016**, *3*, 49. [CrossRef]
10. Kleebusch, E.; Patzig, C.; Krause, M.; Hu, Y.; Höche, T.; Rüssel, C. The formation of nanocrystalline ZrO_2 nuclei in a Li_2O-Al_2O_3-SiO_2 glass—A combined XANES and TEM study. *Sci. Rep.* **2017**, *7*, 10869. [CrossRef]
11. Zandona, A.; Patzig, C.; Rüdinger, B.; Hochrein, O.; Deubener, J. TiO_2(B) nanocrystals in Ti-doped lithium aluminosilicate glasses. *J. Non-Cryst. Solids* **2019**, *2*, 100025. [CrossRef]
12. Dymshits, O.; Bachina, A.; Alekseeva, I.; Golubkov, V.; Tsenter, M.; Zapalova, S.; Bogdanov, K.; Danilovich, D.; Zhilin, A. Phase transformations upon formation of transparent lithium alumosilicate glass-ceramics nucleated by yttrium niobates. *Ceramics* **2023**, *6*, 1490–1507. [CrossRef]
13. Vigier, M.; Deniard, P.; Gautron, E.; Gautier, N.; Genevois, C.; Ory, S.; Allix, M.; Kacem, I.B.; Jobic, S. Microstructural insights on lithium aluminum silicate (LAS) glass ceramics. *Ceram. Int.* **2024**, *50*, 29011–29015. [CrossRef]
14. Guo, Y.; Wang, J.; Ruan, J.; Han, J.; Xie, J.; Liu, C. Microstructure and ion-exchange properties of glass-ceramics containing $ZnAl_2O_4$ and β-quartz solid solution nanocrystals. *J. Eur. Ceram. Soc.* **2021**, *41*, 5331–5340. [CrossRef]
15. Beall, G.H.; Pinckney, L.R. Nanophase glass-ceramics. *J. Am. Ceram. Soc.* **1999**, *82*, 5–16. [CrossRef]
16. Doherty, P.D. Direct observation of the crystallization of Li_2O-Al_2O_3-SiO_2 glasses containing TiO_2. *J. Am. Ceram. Soc.* **1967**, *50*, 77–80. [CrossRef]
17. Alekseeva, I.P.; Belyaevskaya, N.M.; Bobovich, Y.S.; Tsenter, M.Y.; Chuvaeva, T.I. Recording, interpretation, and some examples of application of Raman spectra for glass ceramics activate with titanium (IV) oxide. *Opt. Spectrosc.* **1978**, *45*, 927–936.
18. Khodakovskaya, R.Y. *Chemistry of Titanium-Containing Glasses and Glass Ceramics*; Chemistry: Moscow, Russia, 1978; pp. 190–198. (In Russian)
19. Dymshits, O.S.; Zhilin, A.A.; Petrov, V.I.; Tsenter, M.Y.; Chuvaeva, T.I.; Golubkov, V.V. A Raman spectroscopic study of phase transformations in titanium-containing lithium aluminosilicate glasses. *Glass Phys. Chem.* **1998**, *24*, 79–96.
20. Alekseeva, I.; Dymshits, O.; Tsenter, M.; Zhilin, A.; Golubkov, V.; Denisov, I.; Skoptsov, N.; Malyarevich, A.; Yumashev, K. Optical applications of glass-ceramics. *J. Non-Cryst. Solids* **2010**, *356*, 3042–3058. [CrossRef]
21. Kleebusch, E.; Patzig, C.; Krause, M.; Hu, Y.; Höche, T.; Rüssel, C. The effect of TiO_2 on nucleation and crystallization of a Li_2O-Al_2O_3-SiO_2 glass investigated by XANES and STEM. *Sci. Rep.* **2018**, *8*, 2929. [CrossRef]
22. Kleebusch, E.; Patzig, C.; Höche, T.; Rüssel, C. The evidence of phase separation droplets in the crystallization process of a Li_2O-Al_2O_3-SiO_2 glass with TiO_2 as nucleating agent—An X-ray diffraction and (S)TEM-study supported by EDX-analysis. *Ceram. Int.* **2018**, *44*, 2919–2926. [CrossRef]
23. Johnston, W.D. Oxidation-reduction equilibria in molten $Na_2O \cdot 2SiO_2$ glass. *J. Am. Ceram. Soc.* **1965**, *48*, 184–190. [CrossRef]
24. Rawal, B.S.; MacCrone, R.K. Optical absorption in an equimolar barium borosilicate glass containing titanium Ions. *J. Non-Cryst. Solids* **1978**, *28*, 337–345. [CrossRef]
25. Kurkjian, C.R.; Peterson, G.E. An EPR study of Ti^{3+}+-Ti^{4+} in TiO_2-SiO_2 glasses. *Phys. Chem. Glas.* **1974**, *15*, 12–17.
26. Morinaga, K.; Yoshida, H.; Takebe, H. Compositional dependence of absorption spectra of Ti^{3+} in silicate, borate, and phosphate glasses. *J. Am. Ceram. Soc.* **1994**, *77*, 3113–3118. [CrossRef]
27. El-Shafi, N.A.; Morsi, M.M. Optical absorption and infrared studies of some silicate glasses containing titanium. *J. Mater. Sci.* **1997**, *32*, 5185–5189. [CrossRef]

28. Andrade, L.H.C.; Lima, S.M.; Novatski, A.; Neto, A.M.; Bento, A.C.; Baesso, M.L.; Gandra, F.C.G.; Guyot, Y.; Boulon, G. Spectroscopic assignments of Ti^{3+} and Ti^{4+} in titanium-doped OH– free low-silica calcium aluminosilicate glass and role of structural defects on the observed long lifetime and high fluorescence of Ti^{3+} ions. *Phys. Rev. B* **2008**, *78*, 224202. [CrossRef]
29. Vogel, W. *Glass Chemistry*; Springer: Berlin/Heidelberg, Germany; New York, NY, USA, 1994; pp. 313–314. [CrossRef]
30. Höland, W.; Wange, P.; Naumann, K.; Vogel, J.; Carl, G.; Jana, C.; Götz, W. Control of phase formation processes in glass-ceramics for medicine and technology. *J. Non-Cryst. Solids* **1991**, *129*, 152–162. [CrossRef]
31. Wisniewski, W.; Thieme, K.; Rüssel, C. Fresnoite glass-ceramics—A review. *Prog. Mater. Sci.* **2018**, *98*, 68–107. [CrossRef]
32. Keding, R.; Rüssel, C. Electrochemical nucleation for the preparation of oriented glass ceramics. *J. Non-Cryst. Solids* **1997**, *219*, 136–141. [CrossRef]
33. Höche, T.; Kleebe, H.-J.; Brydson, R. Can fresnoite ($Ba_2TiSi_2O_8$) incorporate Ti^{3+} when crystallizing from highly reduced melts? *Philos. Mag. A* **2001**, *81*, 825–839. [CrossRef]
34. Eremeev, K.; Dymshits, O.; Alekseeva, I.; Khubetsov, A.; Zapalova, S.; Tsenter, M.; Basyrova, L.; Serres, J.M.; Mateos, X.; Loiko, P.; et al. Effect of redox conditions of glass melting on the structure and the properties of titanium-containing gahnite glass-ceramics. *J. Eur. Ceram. Soc.* **2024**, *44*, 3362–3380. [CrossRef]
35. Lipatiev, A.; Fedotov, S.; Lotarev, S.; Naumov, A.; Lipateva, T.; Savinkov, V.; Shakhgildyan, G.; Sigaev, V. Direct laser writing of depressed-cladding waveguides in extremely low expansion lithium aluminosilicate glass-ceramics. *Opt. Laser Technol.* **2021**, *138*, 106846. [CrossRef]
36. Lipson, H.; Steeple, H. *Interpretation of X-Ray Powder Patterns*; McMillan: London, UK; New York, NY, USA, 1970; p. 344.
37. Schneider, C.A.; Rasband, W.S.; Eliceiri, K.W. NIH Image to ImageJ: 25 years of image analysis. *Nat. Methods* **2012**, *9*, 671–675. [CrossRef] [PubMed]
38. Ovono Ovono, D.; Berre, S.; Pradeau, P.; Comte, M.; Bruno, G. Study of the crystallization kinetics of LAS glass by differential scanning calorimetry, X-ray diffraction, and beam bending viscometry. *Thermochim. Acta* **2012**, *527*, 158–164. [CrossRef]
39. Ovono Ovono, D.; Bruno, G.; Pradeau, P.; Berre, S. Conditions for crystallization of LAS glass-ceramics as a function of nucleating agent amount and heat treatment. *Int. J. Appl. Glass Sci.* **2013**, *4*, 20–30. [CrossRef]
40. Shirasuka, K.; Yanagida, H.; Yamaguchi, G. The preparation of η-alumina and its structure. *Mater. Sci.* **1976**, *84*, 610–613. [CrossRef]
41. Zhou, R.-S.; Snyder, R.L. Structures and transformation mechanisms of the η, γ and θ transition aluminas. *Acta Cryst.* **1991**, *B47*, 617–630. [CrossRef]
42. Skala, R.D.; Li, D.; Low, I.M. Diffraction, structure and phase stability studies on aluminium titanate. *J. Eur. Ceram. Soc.* **2009**, *29*, 67–75. [CrossRef]
43. Galeeener, F.L.; Leadbetter, A.J.; Stringfellow, M.W. Comparison of the neutron, Raman and infrared vibrational spectra of vitreous SiO_2, GeO_2 and BeF_2. *Phys. Rev. B* **1983**, *27*, 1052–1078. [CrossRef]
44. Bobovich, Y.S. Spectroscopic invistigation of titanium coordination in some vitreous materials. *Opt. Spectrosc.* **1963**, *14*, 647–654.
45. Thomas, P.V.; Ramakrishnan, V.; Vaidyan, V.K. Oxidation studies of aluminum thin films by Raman spectroscopy. *Thin Solid Film.* **1989**, *170*, 35–40. [CrossRef]
46. Baronskiy, M.; Rastorguev, A.; Zhuzhgov, A.; Kostyukov, A.; Krivoruchko, O.; Snytnikov, V. Photoluminescence and Raman spectroscopy studies of low-temperature γ-Al_2O_3 phases synthesized from different precursors. *Opt. Mater.* **2016**, *53*, 87–93. [CrossRef]
47. Liu, Y.; Cheng, B.; Wang, K.; Ling, G.; Cai, J.; Song, C.; Han, G. Study of Raman spectra for γ-Al_2O_3 models by using first-principles method. *Solid State Commun.* **2014**, *178*, 16–22. [CrossRef]
48. Basyrova, L.; Bukina, V.; Balabanov, S.; Belyaev, A.; Drobotenko, V.; Dymshits, O.; Alekseeva, I.; Tsenter, M.; Zapalova, S.; Khubetsov, A.; et al. Synthesis, structure and spectroscopy of Fe^{2+}: $MgAl_2O_4$ transparent ceramics and glass-ceramics. *J. Lumin.* **2021**, *236*, 118090. [CrossRef]
49. Sharma, K.; Simons, B. Raman study of crystalline polymorphs and glasses of spodumene composition quenched from various pressures. *Am. Mineral.* **1981**, *66*, 118–126.
50. Alekseeva, I.; Dymshits, O.; Ermakov, V.; Zhilin, A.; Petrov, V.; Tsenter, M. Raman spectroscopy quantifying the composition of stuffed β quartz derivative phases in lithium aluminosilicate glass-ceramics. *J. Non-Cryst. Solids* **2008**, *354*, 4932–4939. [CrossRef]
51. Sprengard, R.; Binder, K.; Brandle, M.; Fotheringham, U.; Sauer, J.; Pannhorst, W. On the interpretation of the experimental Raman spectrum of β-eucryptite $LiAlSiO_4$ from atomistic computer modeling. *J. Non-Cryst. Solids* **2000**, *274*, 264–270. [CrossRef]
52. Bost, N.; Duraipandian, S.; Guimbretière, G.; Poirier, J. Raman spectra of synthetic and natural mullite. *Vib. Spectrosc.* **2016**, *82*, 50–52. [CrossRef]
53. Chavoutier, M.; Caurant, D.; Majérus, O.; Boulesteix, R.; Loiseau, P.; Jousseaume, C.; Brunet, E.; Lecomte, E. Effect of TiO_2 content on the crystallization and the color of (ZrO_2, TiO_2)-doped Li_2O-Al_2O_3-SiO_2 glasses. *J. Non-Cryst. Solids* **2014**, *384*, 15–24. [CrossRef]

54. Loeffler, B.M.; Burns, R.G.; Tossell, J.A.; Vaughan, D.J.; Johnson, K.H. Charge transfer in lunar materials: Interpretation of ultraviolet-visible spectral properties of the moon. In Proceedings of the Lunar Science Conference, 5th, Houston, TX, USA, 18–22 March 1974; Volume 5, pp. 3007–3016.
55. Marfunin, A.S. *Physics of Minerals and Inorganic Materials: An Introduction*; Springer-Verlag: Berlin/Heidelberg, Germany; New York, NY, USA, 1979; p. 340.
56. Petrovski, G.T.; Zhilin, A.A.; Shepilov, M.P.; B'en, V.B.; Li, K.K. Nonmonotonic transmittance variation of a material during the crystallization of liquating glasses. *J. Opt. Technol.* **2003**, *70*, 857–863. [CrossRef]
57. Sanchez, A.; Strauss, A.J.; Aggarwal, R.L.; Fahey, R.E. Crystal Growth, Spectroscopy, and laser characteristics of Ti: Al_2O_3. *IEEE J. Quantum Electron.* **1988**, *24*, 995–1002. [CrossRef]
58. Ripenko, V.V.; Khomenko, V.M.; Vyshnevskyi, O.A.; Kosorukov, O.O. Synthesis and physical properties of Li-Al-Si glass-ceramic materials from petalite of Polohivka Deposit (the Ukrainian Shield). *Mineral. J.* **2017**, *39*, 53–62. [CrossRef]
59. Moulton, P.F.; Cederberg, J.G.; Stevens, K.T.; Foundos, G.; Koselja, M.; Preclikova, J. Characterization of absorption bands in Ti:sapphire crystals. *Opt. Mater. Express.* **2019**, *9*, 2216–2251. [CrossRef]
60. Mohapatra, S.K.; Kröger, F.A. Defect structure of α-Al_2O_3 doped with titanium. *J. Am. Ceram. Soc.* **1977**, *60*, 381–387. [CrossRef]

materials

MDPI

Article

Microstructure and Bioactivity of Ca- and Mg-Modified Silicon Oxycarbide-Based Amorphous Ceramics

Qidong Liu [1], Hongmei Chen [2], Xiumei Wu [1], Junjie Yan [1], Biaobiao Yang [3,4,5], Chenying Shi [6,*], Yunping Li [3] and Shu Yu [1,*]

1 National Key Laboratory of Science and Technology on High-Strength Structural Materials, Central South University, Changsha 410083, China; wxiumei@csu.edu.cn (X.W.); matti28jy@gmail.com (J.Y.)
2 School of Materials Science and Engineering, Harbin Institute of Technology (Shenzhen), Shenzhen 518055, China
3 State Key Lab for Powder Metallurgy, Central South University, Changsha 410083, China; biaobiao.yang@imdea.org (B.Y.)
4 IMDEA Materials Institute, C/Eric Kandel 2, Getafe, 28906 Madrid, Spain
5 Department of Materials Science, Polytechnic University of Madrid/Universidad Politécnica de Madrid, E.T.S. de Ingenieros de Caminos, 28040 Madrid, Spain
6 College of Chemistry and Chemical Engineering, Central South University, Changsha 410083, China
* Correspondence: 22023079@csu.edu.cn (C.S.); yushu@csu.edu.cn (S.Y.)

Abstract: Silicon oxycarbide (SiOC), Ca- and Mg-modified silicon oxycarbide (SiCaOC and SiMgOC) were synthesized via sol–gel processing with subsequent pyrolysis in an inert gas atmosphere. The physicochemical structures of the materials were characterized by XRD, SEM, FTIR, and ^{29}Si MAS NMR. Biocompatibility and in vitro bioactivity were detected by MTT, cell adhesion assay, and simulated body fluid (SBF) immersion test. Mg and Ca were successfully doped into the network structure of SiOC, and the non-bridging oxygens (NBO) were formed. The hydroxycarbonate apatite (HCA) was formed on the modified SiOC surface after soaking in simulated body fluid (SBF) for 14 days, and the HCA generation rate of SiCaOC was higher than that of SiMgOC. Accompanying the increase of bioactivity, the network connectivity (NC) of the modified SiOC decreased from 6.05 of SiOC to 5.80 of SiCaOC and 5.60 of SiMgOC. However, structural characterization and biological experiments revealed the nonlinear relationship between the biological activity and NC of the modified SiOC materials.

Keywords: sol–gel; SiOC; network connectivity; apatite forming ability; cytotoxicity

1. Introduction

Bioactive glasses (BGs) have been widely used in bone regeneration and tissue engineering due to their good bioactivity, biocompatibility, excellent chemical stability, and well-defined bonding ability with soft tissues and bone, as well as high bone regeneration capacity [1,2]. The superior bioactivity of BGs is greatly attributed to the formation of the bone-like hydroxycarbonate apatite (HCA, $Ca_{10}(PO_4)_6(OH)_2 \cdot CO_3$) layer in the biological environment, inducing strong bonding of the BG surface to the bone tissue [3]. Rawlings [4] and Strnad [5] proposed the network connectivity (NC) of bioactive glasses and revealed the relationship between NC and bioactivity. To be specific, the tetrahedrons were described by the notation Q^n, in which n denoted the number of bridging oxygen atoms in the SiO_4 unit and varied between zero and four [6].

However, two shortcomings limited the application of BGs: the poor mechanical strength and the inability to resist crystallization during the preparation [7,8]. Recently, silicon oxycarbide (SiOC)-based materials have been used in biomedical applications for bone defect repair, such as load-supporting implants or biological coatings [9]. The surface reactivity of SiOC-based materials is similar to that of BGs [10], and they can resist crystallization under higher temperatures (≤1300 °C) [11]. This can be credited to the unique microstructure of SiOC, which has a network of corner-shared silicon-centered tetrahedrons containing both Si–C and Si–O bonds but no C–O bonds [12]. SiOC exhibits the full range of mixed bonded SiO_xC_{4-x} tetrahedra, i.e., SiO_4, SiO_3C, SiO_2C_2, $SiOC_3$, and SiC_4, as well as a few free carbons [13,14].

Although SiOC exhibits better biocompatibility, it is biologically inert. Doping the SiOC glass network with other elements, especially alkaline earth metals such as Li, Ca, and Mg, can increase biological activity and reduce its NC [15,16]. Calcium (Ca) plays a crucial role in bone formation and resorption as the primary constituent of biological apatite ($Ca_{10}(PO_4)_6(OH)_2$) [17]. Magnesium (Mg) as an abundant and essential mineral in the human body, can inhibit osteoclasts and enhance osteoblast activity [18]. In general, the addition of Ca and Mg was found to enhance the biocompatibility and osteoinduction of the materials and promote cell adhesion [19].For instance, Xie et al. [20] synthesized Ca and/or B-modified SiOC and investigated their crystallization behavior, network architecture, and chemical compositions. Ca/B doping could decrease the NC of SiOC from 5.95 to 5–5.5. However, their investigation was unfortunately limited to the structural evolutions in the material without further delving into the influence of doping on the bioactivity and biocompatibility of SiOC. Chen et al. [21] reported the excellent biocompatibility and bioactivity of SiCaMgOC coatings roughly but showed that they lacked quantitative evaluation of biological activity using advanced techniques as well as a critical understanding regarding the influence of individual elements (Mg or Ca) on the microstructure and bioactivity. Ionescu et al. [22] demonstrated a weak negative correlation between the biological activity and NC of modified SiOC by alkaline earth elements (Mg and Ca), and the quantitative model for glasses could be obtained by calculating the number of bridging and non-bridging oxygens (NBO) per silicon–oxygen tetrahedron in modified SiOC. But, their work in biological activity was only evaluated by immersion in simulated body fluid (SBF) to induce bone-like HAP without incorporating it with other important cell compatibility experiments.

There are very few reports on the modification research of SiOC-based materials, and there is no systematic correlation between material structure and biological activity. The objective of this study is to examine the bioactivity of Ca- and Mg-modified SiOC and establish the quantifiable evaluation of material bioactivity through the changes of NC. The quantitative structure–bioactivity model through computational algorithms in materials science will also be considered in future work [23], and the highly bioactive SiOC will be applied in bone defect repair.

To this end, the microstructures and bioactivity of Ca- and Mg-modified SiOC amorphous ceramics were investigated thoroughly in comparison with SiOC without any doping. Previous studies demonstrated that with increasing concentrations of Ca and Mg, the bioactivity of the material was enhanced while the biocompatibility decreased. Therefore, we selected the samples with the highest biocompatibility for subsequent experiments. The modified SiOCs doped with Ca and Mg were prepared by the sol–gel method, using dimethyldiethoxysilane and methyltriethoxysilane as precursors and Ca/Mg acetylacetonate as modifiers. The biological activity was evaluated through inducing HCA in SBF at 37 °C. In addition, the biocompatibility of the materials was investigated by MTT assay and cell adhesion assay with mouse fibroblast L929 cells.

2. Materials and Methods

2.1. Materials Preparation

To prepare the Ca/Mg modified SiOC glasses, 1 mol dimethyldiethoxysilane (DMDES, AR, 98%, Aladdin, Shanghai, China) and 1 mol methyltriethoxysilane (MTES, AR, 98%, Aladdin, Shanghai, China) were mixed as precursors, which were further mixed with deionized water and anhydrous ethanol to keep the molar ratio of H_2O/OCH_3^- for 1 and EtOH/Si for 2. After magnetic stirring in a water bath at 40 °C for 30 min, 0.05 mol of Ca acetylacetonate ($Ca(acac)_2$, AR, 98%, Aladdin, Shanghai, China) or 0.06 mol of Mg acetylacetonate ($Mg(acac)_2$, AR, 98%, Aladdin, Shanghai, China) were added to hydrolysis for 90 min. Note that the Ca:Si molar ratio is 1:40, and Mg:Si molar ratio is 1:33. Hydrochloric acid was added to keep pH at 4. The aged gel can be kept at room temperature for one week.

C/C substrates (φ 10 × 3 mm, 1.75 g/cm^3) were manually ground with 400#, 1000#, and 2000# sandpapers, then corroded with aqua regia for 30 min, and cleaned with anhydrous ethanol and deionized water successively. The substrates were coated with aged gel with a lift-off speed of 20 mm/min, then dried at 80 °C, and finally, subjected to pyrolysis in argon gas at 1000 °C to obtain the ceramic coating. The pyrolyzed samples with and without Ca/Mg modified SiOC ceramic coating were named SiCaOC, SiMgOC, and SiOC, respectively.

2.2. Materials Characterization

The phase constitution of the samples was determined by X-ray diffraction (XRD, Advance D8, Bruker, Romanshorn, Switzerland) equipped with Cu-Kα X-ray source in 2θ range from 10° to 60° with 6°/min of scanning rate. The surface morphology of samples was observed by SEM (Nova NanoSEM230, FEI, Prague, Czech Republic) in the secondary electron mode with a vacuum degree below 5×10^{-5} Pa and an acceleration voltage of 5 kV. The absorption spectra of the materials were recorded through FTIR (Spectrum Two with STA8000, PerkinElmer, Eindhoven, The Netherlands). The microstructures were characterized using ^{29}Si MAS NMR. ^{29}Si spectra were recorded utilizing Bruker ZG sequence at spinning rates of 8 kHz, pulse angles of 22°, and 120 s relaxation delay. The carbon content of the samples was detected by a CS 600 carbon analyzer (Leco Corporation, St. Joseph, MI, USA), and the oxygen content was detected by a TCH 600 Nitrogen/Oxygen/Hydrogen analyzer (Leco Corporation, MI, USA). In addition, Si, Ca, and Mg contents in SBF were detected by Inductively Coupled Plasma Optical Emission Spectrometry (ICP-OES, TCP-5100-VDV, NYSE: A, Santa Clara, CA, USA).

2.3. Mineralization in SBF

Simulated body fluid (SBF) was prepared according to Kokubo's [24] method to evaluate the vitro acellular bioactivity. During this experiment, the samples were placed vertically in the SBF at 37 °C. After the immersion, the samples were successively rinsed with acetone (AC, AR, 98%, Aladdin, Shanghai, China), deionized water, and anhydrous ethanol and then dried at 40 °C. The content of release ions was detected by ICP-OES, and the pH of SBF was measured by a pH analyzer (TP110, TIMEPOWER, Shanghai, China), respectively. The microstructures of samples after immersion in SBF were further observed by SEM.

2.4. Cytotoxicity Test

MTT (3-[4,5-dimethylthiazol-2-yl]-2,5-diphenyltetrazolium bromide) assay was used to evaluate the L929 cells' proliferation after in contact with different materials extracts, based on the reduction in tetrazolium salt by living cells [25,26]. To be specific, SiCaOC, SiMgOC, and SiOC were sterilized by the autoclave (LDZM-100KBS, Shanghai, China) and then immersed in the complete medium for 24 h, 48 h, and 72 h to obtain extracts of different concentrations. Materials extracts and Dulcecco's modified eagle medium (DMEM) were used as experimental groups and the negative control group, respectively.

L929 cells were cultured in DMEM supplemented with 10% fetal calf serum (FCS) at 37 °C and 5% CO_2. Then, 100 μL cell suspension with a density of 5×10^4 cells/mL was seeded onto the bottom of the 96-well cell culture plate and continuously cultured for 24 h. After 24 h of cell culture, 100 μL material extract or 100 μL DMEM was added to cell suspension corresponding to the experimental groups and negative control group, respectively. The cells were incubated at 37 °C and 5% CO_2 for 72 h. After that, 10 μL of 5 mg/mL MTT was added to all groups. The cells were incubated in the mixture of MTT and DMEM for 4 h. After that, 150 μL dimethyl sulfoxide (DMSO) was added to dissolve the cells. The optical absorbance (OD) of the supernatant was measured at a wavelength of

630 nm using the automatic plate reader (ELX800, BioTek, Winooski, VT, USA). The relative growth rate (RGR) of the cells was determined according to the formula in Equation (1) [27]:

$$\text{RGR}(\%) = \frac{\text{OD}(\text{experimental group})}{\text{OD}(\text{negative control})} \times 100\% \quad (1)$$

Cell viability (RGR) of the negative control group measured by the MTT method was defined as 100%. Then, the RGR of the experimental groups was calculated relative to the negative control group. The cytotoxicity of the material is categorized into different grades: RGR $\geq$ 100%; 75~95%; and 50~74% for grades 0, 1, and 2, respectively. Both grades 0 and 1 stand for non-cytotoxicity. Moreover, grade 2 corresponds to non-cytotoxicity or cytotoxicity, depending on the final morphology of cells. The experiment was duplicated three times under the same conditions.

2.5. Cell Adhesion Assay

L929 cell suspension with a density of 5 $\times$ 10^4 cells/mL was dropped evenly on the surface of the three samples. After 72 h incubation, the samples were rinsed with phosphate buffer saline (PBS), then fixed with 2.5% glutaraldehyde solution for 2 h, and finally dehydrated with ethanol. The morphology of the cells adhered to the surface of the materials was further observed by SEM.

2.6. Statistical Analysis

All the analyses were repeated at least three times, and the data were analyzed by GraphPad Prism (ver.9.0.0) software using ANOVA one and two-way followed by Tukey post hoc test to determine the difference significance [28,29]. The significances were shown by *p*-value using * sign.

3. Results and Discussion

3.1. Microstructures

The phase constitutions of Ca-modified and Mg-modified SiOC glasses, as well as SiOC without any doping, were analyzed by XRD, as shown in Figure 1. The incorporation of Ca or Mg did not change the phase constitution of SiOC, i.e., amorphous SiO_2 corresponding to a broad peak between 20° and 25°, which indicated that the three materials were completely amorphous after 1000 °C pyrolysis.

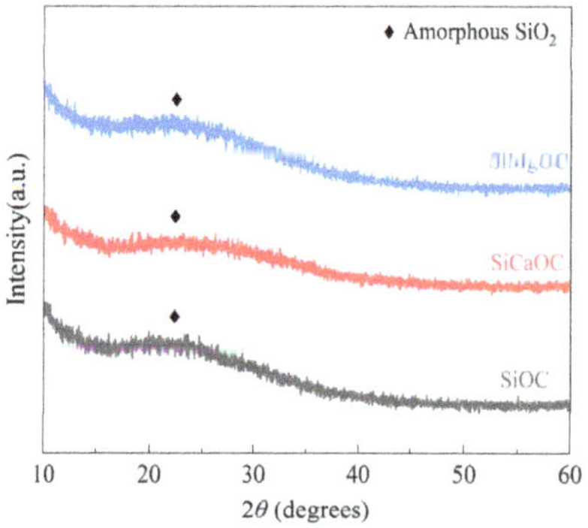

Figure 1. XRD analysis of different samples after 1000 °C heat treatment.

Surface morphology and elemental distribution of initial and modified SiOC are shown in Figure 2. The surface of SiOC was relatively smooth, with some minor cracks caused by thermal stress during the pyrolysis process. In comparison with SiOC, the SiCaOC had a much rougher surface with some obvious puddles, while the SiMgOC displayed a rougher surface with some pronounced cracks. SiCaOC and SiMgOC showed fewer cracks, maybe because the rougher surfaces were in favor of stress release. Further EDS results evidenced that both Ca and Mg have been distributed uniformly on the surface of glass materials.

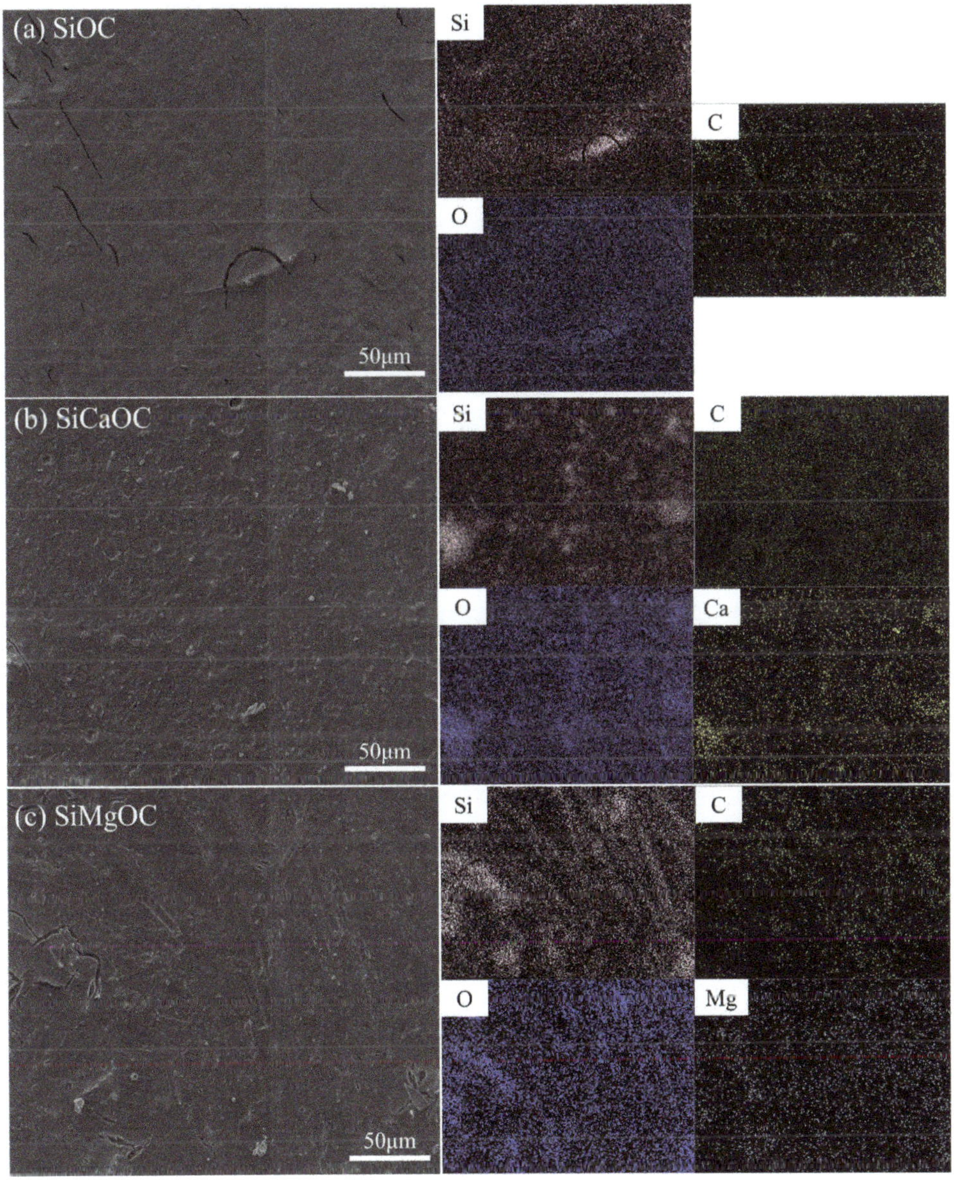

Figure 2. SEM and EDS analysis of different materials: (**a**) SiOC glass; (**b**) SiCaOC glass; (**c**) SiMgOC glass.

The chemical bands of three glass samples were investigated by FTIR spectroscopy, as shown in Figure 3. Both the absorptions at around 1083 and 1220 cm^{-1} correspond to the asymmetric stretching vibrations of the Si–O–Si [20,30]. The absorption at 800 cm^{-1} was overlapped by symmetric stretching and bending vibration of Si–O–Si, as well as bending vibration of Si–C. According to a study by Lyu et al. [31], the modification of Ca and Hf into the SiOC can form silica sites with NBO (Si–O$^-$), leading to the appearance of absorption in the range of 900–970 cm^{-1}. In this manner, the new absorptions appearing at 939 cm^{-1} and 957 cm^{-1} could be explained by the presence of NBO in SiO_4 tetrahedral units [32]. This also indicates the successful fabrication of SiCaOC and SiMgOC.

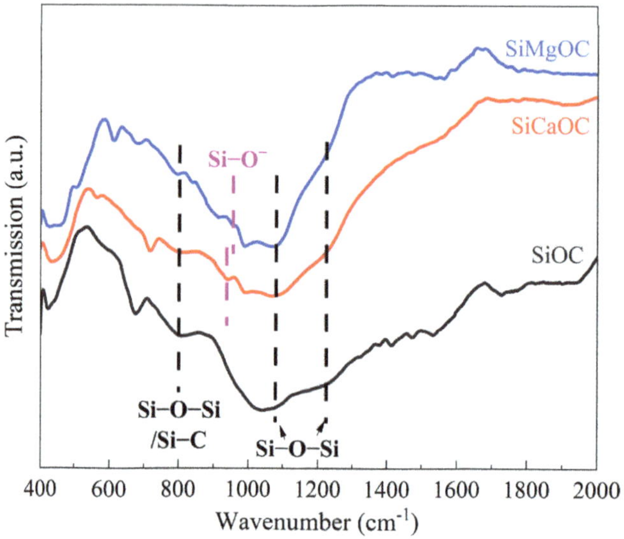

Figure 3. FTIR spectra of SiOC, SiCaOC, and SiMgOC glasses.

The microstructure and network architecture of samples were investigated by ^{29}Si MAS NMR spectra (Figure 4). In addition, based on the experimental NMR data, the values of NC for various glass samples were determined from the following equation [22]:

$$NC = [4 \times f(Q^4)] + [3 \times f(Q^3)] + [2 \times f(Q^2)] + [1 \times f(Q^1)] + [6 \times f(SiO_3C)] + [8 \times f(SiO_2C_2)] + [10 \times f(SiOC_3)] + [12 \times f(SiC_4)] \quad (2)$$

with Q being the Si atomic fractions among total network formers and $f(x)$ the site fraction of the corresponding species x from the experimental NMR data. The NC^f of the samples was calculated by Equation (2) and summarized in Table 1.

Table 1. Silicon site fractions of three SiOC-based glasses determined from ^{29}Si–MAS NMR spectra.

Sample	Q^4SiO_4 (I)	Q^3SiO_4 (II)	SiO_3C (III)	SiO_2C_2 (IV)	$SiOC_3$ (V)	SiC_4 (VI)	NMR Composition	Measured Composition	NC
SiOC	42.1%	–	30.6%	18.6%	–	8.7%	$Si_1O_{1.49}C_{0.25}$	$Si_1O_{1.49}C_{0.51}$	6.05
SiCaOC	51.6%	1.8%	21.3%	11.1%	9.2%	5.0%	$Si_1Ca_{0.05}O_{1.54}C_{0.23}$	$Si_1Ca_{0.05}O_{1.54}C_{0.56}$	5.80
SiMgOC	56.8%	2.8%	18.4%	8.4%	7.9%	5.7%	$Si_1Mg_{0.06}O_{1.59}C_{0.20}$	$Si_1Mg_{0.06}O_{1.59}C_{0.57}$	5.60

The spectra of SiOC in Figure 4a were fitted with Gaussian peaks at around −107, −71, −34, and −7 ppm, which correspond to the Q_4SiO_4(I), SiO_3C (III), SiO_2C_2 (IV), and SiC_4 (VI), respectively [12]. Furthermore, two new peaks at −91 and −11 ppm of both SiCaOC and SiMgOC corresponded with Q_3SiO_4 (II) and $SiOC_3$ (V), respectively, and those peaks were hardly observed in the spectra of SiOC (Figure 4).

The relative fractions of different silicon species in three studied glass materials that were analyzed from NMR spectra are shown in Table 1. Additionally, the relative fractions of various SiOC-based glasses were also determined by carbon analyzer, nitrogen/oxygen/hydrogen analyzer, and ICP-OES, whose results were roughly in line with the NMR composition except the higher content of carbon for the free carbon existed in SiOC glass, rather than in the SiOC network (Table 1).

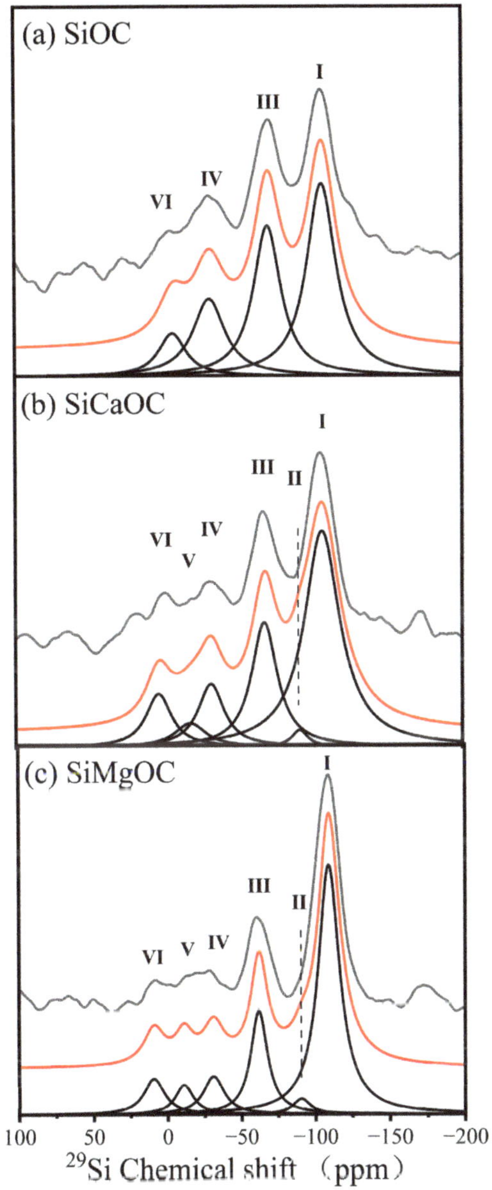

Figure 4. ^{29}Si MAS NMR of three materials: (**a**) SiOC glass; (**b**) SiCaOC glass; (**c**) SiMgOC glass. The experimental (grey line) and simulated (red line) spectra, as well as the individual simulation components (black lines), are shown. The results of the simulation correspond to Q_4SiO_4 (I), Q_3SiO_4 (II), SiO_3C (III), SiO_2C_2 (IV), $SiOC_3$ (V), and SiC_4 (VI).

The proportion of Q_4SiO_4 sites increased significantly from 42.1% for SiOC to 51.6% for SiCaOC and 56.8% for SiMgOC (Table 1). Accordingly, the fractions of SiO_3C, SiO_2C_2,

$SiOC_3$, and SiC_4 sites decreased after modification. Moreover, the addition of alkaline earth metals results in the formation of a small fraction of Q_3SiO_4 sites, i.e., 1.8% for SiCaOC and 2.8% for SiMgOC. The NMR composition showed that the O/Si ratio increased from 1.49 for SiOC to 1.54 for SiCaOC and 1.59 for SiMgOC, while the C/Si ratio decreased from 0.25 for SiOC to 0.23 for SiCaOC and 0.20 for SiMgOC. That means more oxygen-rich Si tetrahedral was formed after doping of Ca or Mg. Moreover, the modification with Ca or Mg reduced the NC of the SiOC-based glasses, from 6.05 for SiOC to 5.80 for SiCaOC and 5.60 for SiMgOC (Table 1).

3.2. *In Vitro Bioactivity*

The apatite-forming ability of SiOC, SiCaOC, and SiMgOC were evaluated by immersion in SBF, and their surface morphologies after different immersion periods were observed by SEM, as shown in Figure 5. The surface of SiOC remained unchanged even after the immersion in SBF for 14 days (Figure 5a–d). However, clear spherical sediments with a diameter of 1–3 μm were observed on the surface of SiCaOC after immersion in SBF for 3 days (Figure 5e). As the immersion time prolonged to 7 days, these sediments were observed to grow up with a diameter of 6–10 μm (Figure 5f). At an immersion time longer than 10 days, the surface of SiCaOC was coated fully with these sediments (Figure 5g–h). The surface evolution of SiMgOC in SBF was similar to that of SiCaOC but had two differences. The first one was that the appearance of sediments in SiMgOC seemed to be postponed since no sediments formed at 3 days of immersion (Figure 5i) and limited sediments appeared at 7 days of immersion (Figure 5j) with almost full coverage of sediments until 14 days of immersion (Figure 5l). This demonstrates that SiCaOC has a higher ability to generate sediments than SiMgOC. The other difference is in the larger size of sediments in SiMgOC (diameter: 5–15 μm) compared to that of SiCaOC.

The morphologies of sediments at higher magnifications and corresponding EDS analysis for SiCaOC and SiMgOC are shown in Figure 6. Both sediments exhibit a spherical cauliflower morphology, which is consistent with previous studies by Haider et al. [33] and Chen et al. [34]. Further EDS analysis revealed that the Ca/P molar ratio of the sediments was 1.70 for SiCaOC and 1.69 for SiMgOC, respectively. Note that both values were almost identical to that of stoichiometric HCA (1.69) [35], whose appearance can also be evidenced by the formation of crystal nuclei in epitaxy on the material surface (Figure 6) [36]. The modified SiOC (SiCaOC or SiMgOC) showed a higher bioactivity than SiOC, given they can form HCA in SBF, and the induction of HCA for SiCaOC was faster than that of SiMgOC. This may be related to the fact that Ca is the necessary element of HCA as well as Mg^{2+}, which can reduce the conversion rate of amorphous calcium–phosphate to the crystalline HCA [37,38].

Figure 7 shows the change in different ion concentrations and pH values of the SBF for 14 days. The Si ion concentration of all materials gradually increased as immersion time increased, while the releasing rate of Si was sensitive to the materials, i.e., SiCaOC > SiMgOC > SiOC (Figure 7a). However, the evolution of Ca ion concentration with respect to the immersion time depends on the material itself. Ca ion of SiCaOC increased sharply to 148.9 mg/L in the first two days for the release of Ca^{2+} from SiCaOC, then decreased drastically to the lowest level (62.7 mg/L) after 14 days among all the materials (Figure 7b). With the immersion time increased, SiMgOC exhibited a continuous decline in Ca ion concentration, reaching the value of 69.3 mg/L after 14 days. A similar decreasing trend was observed for SiOC, while this decreasing rate was much lower in comparison with SiMgOC (Figure 7b). The concentration of P ions in the SBF of all experimental groups exhibited a gradual decrease with increasing immersion time, and the decreasing rate was sensitive to the materials, i.e., SiCaOC > SiMgOC > SiOC (~0) (Figure 7c). Note that the generation of HCA in SBF may explain the decline in Ca and P ion concentrations of SBF during the immersion of SiCaOC and SiMgOC. The pH of SBF immersed by both SiCaOC and SiMgOC increased rapidly but then remained almost constant as immersion time increased, but both were higher than that of SiOC (Figure 7d). The increase in pH for SiCaOC and SiMgOC after initial immersion could be related to the exchange of Ca^{2+} or Mg^{2+} in the materials with H^+ or H_3O^+ in the SBF, accompanied by the formation of Si–OH and the breaking of Si–O–Si in materials [39,40]. The subsequent formation of the HCA layer consumed carbonate and phosphate ions, leading to the formation of more H^+ or H_3O^+ in the solution via shifting the following reversible reactions in the right direction [41]:

$$HCO_3^- \leftrightarrow CO_3^{2+} + H^+ \tag{3}$$

$$HPO_4^{2-} \leftrightarrow PO_4^{3-} + H^+ \tag{4}$$

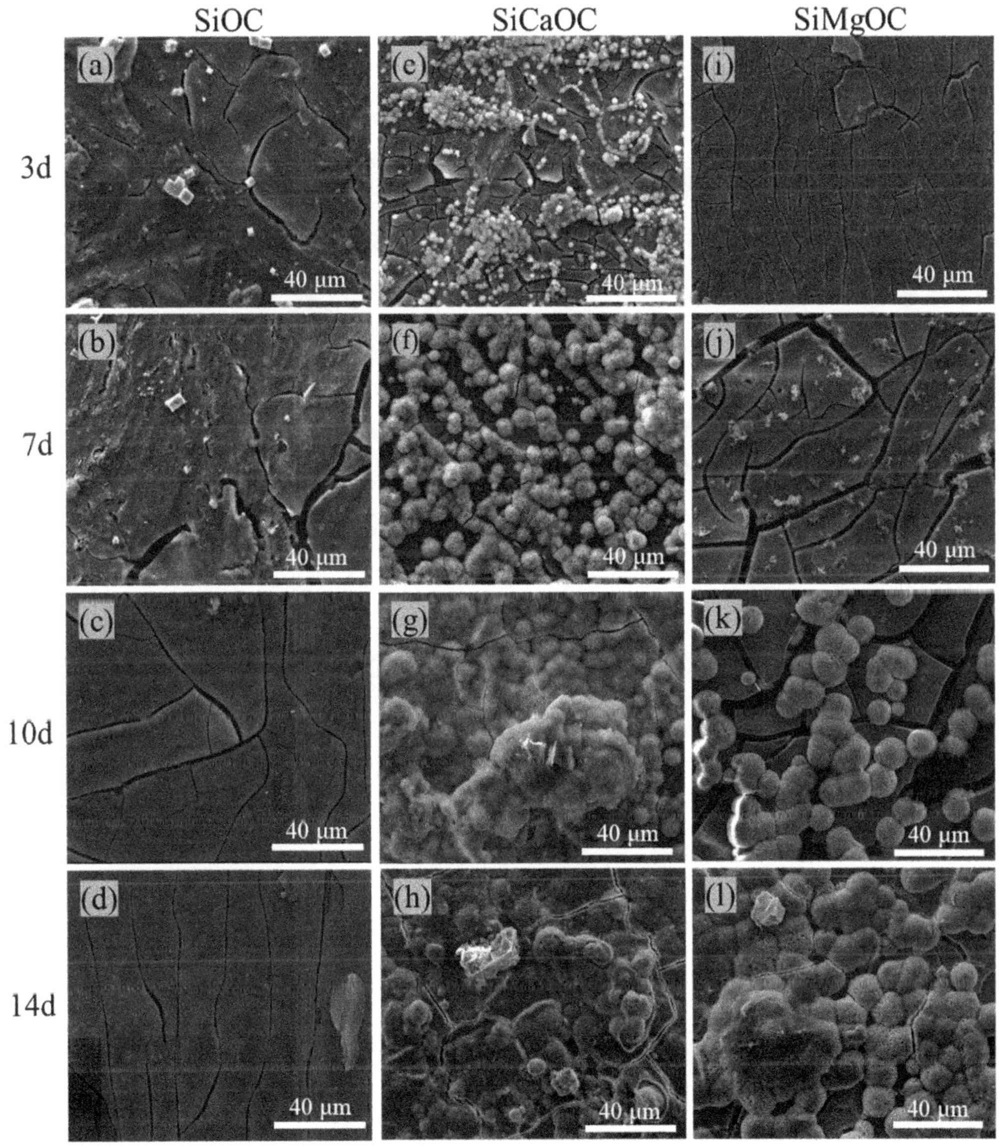

Figure 5. The surface morphology of the samples immersed in SBF for different immersion times: (**a**–**d**) SiOC glass; (**e**–**h**) SiCaOC glass; (**i**–**l**) SiMgOC glass.

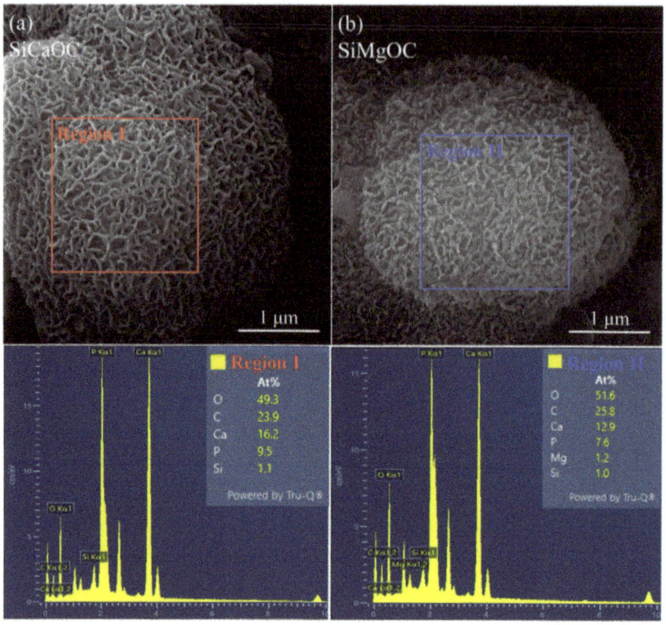

Figure 6. SEM morphology and corresponding EDS spectra of glass materials after 14-day immersion in SBF: (**a**) SiCaOC and (**b**) SiMgOC.

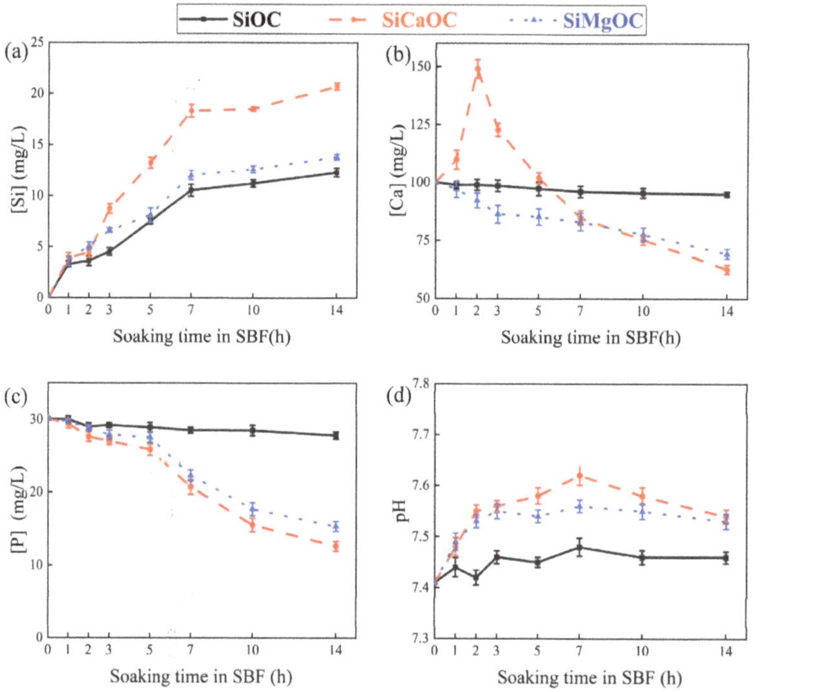

Figure 7. The change in different ions content and pH of SBF after the immersion tests for various glass materials: (**a**) [Si]; (**b**) [Ca]; (**c**) [P]; and (**d**) pH.

Their combination brought the constant value of pH for SBF immersed by SiCaOC and SiMgOC, while the missing of these reactions in SiOC should be responsible for the stable value of pH (Figure 5). In addition, after 14 days, the pH reached ~7.54 for SiCaOC and SiMgOC, slightly higher than that for SiOC (~7.45).

3.3. Biocompatibility Assessment

The cytotoxicity of the studied glass materials was evaluated by MTT assay with L929 cells for culturing for 3 days, with the results shown in Figure 8 and Table 2. With the increase in extraction time, the absorbance of all experimental groups decreased, indicating the number of living cells decreased. The cell viability of SiMgOC was higher than that of the negative control, while the other two experimental groups (SiCaOC and SiOC) were lower than that of the negative control (Figure 8). In addition, the cytotoxicity grade of SiMgOC was 0, while that of SiOC and SiCaOC was 1, which indicated that SiMgOC was the least toxic to cells, followed by SiCaOC and SiOC. It was noteworthy that the cytotoxicity grade of either 0 or 1 indicated non-cytotoxicity [27]. Furthermore, SiMgOC extract can significantly promote cell reproduction and differentiation, especially in low concentrations. The results of the cytotoxicity test were in agreement with those previously reported by Chen et al. [21].

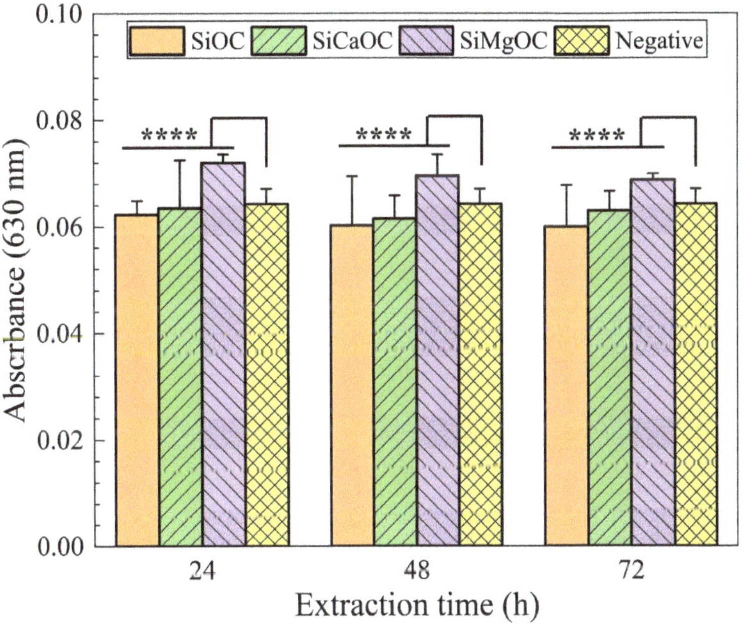

Figure 8. Absorbance of the different samples in MTT toxicity test. n = 3, for each group (**** $p < 0.0001$).

Table 2. The cytotoxicity grades of the materials.

Samples	Grade		
	Extraction of 24 h	**Extraction of 48 h**	**Extraction of 72 h**
SiOC	1	1	1
SiCaOC	1	1	1
SiMgOC	0	0	0
Negative	0	0	0

Figure 9 shows the L929 cell morphology on the surface of the materials with the same culture time of 72 h. The densest cells were observed on the surface of SiCaOC, followed by SiMgOC and SiOC, accompanied by the decreasing bioactivity in Figure 5. Few cells attached to the surface of SiOC grew well with obvious pseudopods (Figure 9a). On the surface of SiCaOC, at least two layers of cells were observed (Figure 9b), causing poor growth or death of cells. However, the surface of the SiMgOC was covered with a layer of cells, and the cells grew well even around the cracks of the surface (Figure 9c). This could explain well why the cytotoxicity of SiCaOC was higher than that of SiMgOC (Table 2), although the bioactivity of SiCaOC was obviously higher than that of SiMgOC (Figure 5).

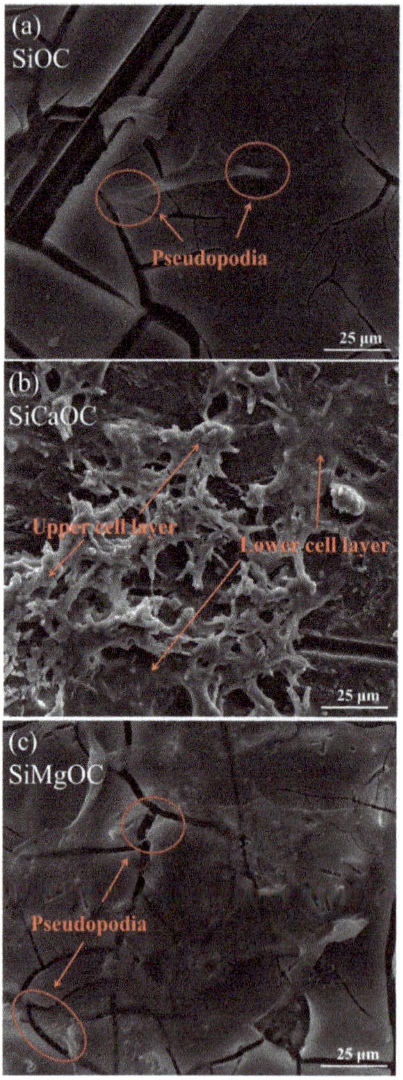

Figure 9. SEM images of L929 cells attached to the surface of the different samples: (**a**) SiOC; (**b**) SiCaOC; and (**c**) SiMgOC.

3.4. Correlation Between Microstructure and Bioactivity

The doping of Ca and Mg resulted in the breakdown of the Si–O–Si bonds in the SiOC network structure, leading to the formation of NBO (Si–O^-), as schematically shown in Figure 10. The Si–O^- does not connect to the other ions directly by "bonding", but rather, it interacts with Ca^{2+} or Mg^{2+} through electrostatic interactions [9,15]. The formation of NBO results in the disruption of the glass network, which, in turn, renders the glass more reactive in aqueous solution like SBF (Figure 5).

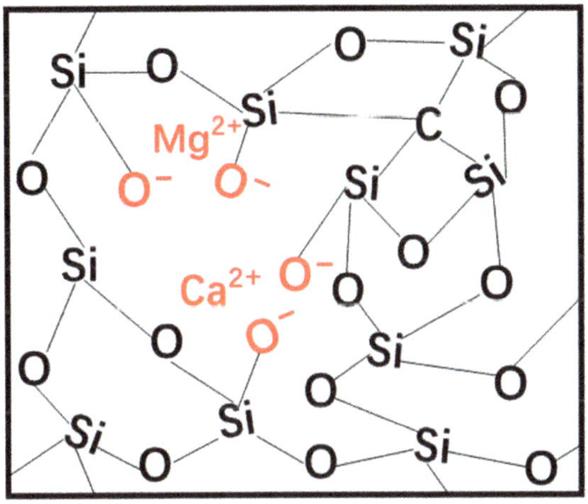

Figure 10. Schematic illustration of the doping of Ca and Mg into the SiOC network structure.

According to the research of Shearer et al. [19], the apatite formation of glass was dependent on the level of polymerization. The higher the NC of the glass, the lower the ion concentration released, which slowed down the formation of apatite. The formation of NBO (Figures 3 and 4) decreased the NC (Table 1) and, thus, improved the glass bioactivity in SBF (Figure 5). However, the NC of SiMgOC was lower than that of SiCaOC, but SiCaOC showed higher bioactivity, as evidenced by faster induction of HCA on the surface (Figure 5). This suggested that the NC may not be the sole determinant of the bioactivity of glass materials, and other microstructural features should be included in designing SiOC-based glasses with excellent bioactivity. In addition, the higher bioactivity still cannot guarantee lower cytotoxicity since multiple layers of cells may be formed on the surface, leading to the constrained growth and premature death of cells (Figure 9). These complicated scenarios for the selection of doping elements should be carefully treated after involving more microstructural factors. The above results demonstrated that the modified SiOC exhibited excellent bioactivity and biocompatibility, thereby indicating its potential for application in the biomedical field of bone defect repair.

4. Conclusions

In this study, SiOC, SiCaOC, and SiCaOC were successfully synthesized by sol–gel method. The microstructure and bioactivity of these three glasses were observed and quantitatively evaluated, respectively. The following conclusions can be derived from this study:

(1) Ca and Mg were successfully doped into the network structure of SiOC by breaking down the Si–O–Si bonding, leading to the formation of NBO. The incorporation of Ca and Mg significantly reduced the NC of SiOC glasses from 6.05 to 5.80 of SiCaOC and 5.60 of SiMgOC, respectively;

(2) The bioactivity of modified materials was obviously improved, as evidenced by the more rapid deposition of HCA on SiCaOC and SiMgOC. Both SiCaOC and SiMgOC

demonstrated excellent biocompatibility, with the cells exhibiting distinct growth on their surface;

(3) Among the three glasses, the highest bioactivity was found in SiCaOC, while the lowest cytotoxicity was found in SiMgOC. Excessive bioactivity will lead to multiple layers of cells on the surface, causing constrained growth as well as premature death of cells;

(4) NC was not the sole determinant of the materials' bioactivity, and other microstructural features should be included in designing high-performance SiOC-based glasses;

(5) The biocompatibility experiments conducted in this study are preliminary and insufficient. Comprehensive evaluations, including cell proliferation, sensitization, and teratogenicity, as well as animal implantation, should be conducted in our next research.

Author Contributions: Conceptualization, Q.L. and S.Y.; Methodology, Q.L. and H.C.; Formal analysis, H.C. and X.W.; Investigation, Q.L., J.Y., B.Y., C.S. and H.C.; Resources, Q.L.; Data curation, Q.L. and H.C.; Writing—original draft, Q.L. and S.Y.; Writing—review & editing, Q.L.; Visualization, Q.L.; Supervision, Q.L.; Project administration, Q.L. and Y.L. All authors have read and agreed to the published version of the manuscript.

Funding: This research received no external funding.

Institutional Review Board Statement: Not applicable.

Informed Consent Statement: Not applicable.

Data Availability Statement: The original contributions presented in the study are included in the article, and further inquiries can be directed to the corresponding author.

Conflicts of Interest: The authors declare no conflict of interest.

References

1. Hench, L.L. The story of Bioglass®. *J. Mater. Sci.-Mater. Med.* **2006**, *17*, 967–978. [CrossRef]
2. Fernandes, H.R.; Gaddam, A.; Rebelo, A.; Brazete, D.; Stan, G.E.; Ferreira, J.M.F. Bioactive Glasses and Glass-Ceramics for Healthcare Applications in Bone Regeneration and Tissue Engineering. *Materials* **2018**, *11*, 2530. [CrossRef]
3. Patel, N.; Best, S.M.; Bonfield, W.; Gibson, I.R.; Hing, K.A.; Damien, E.; Revell, P.A. A comparative study on the in vivo behavior of hydroxyapatite and silicon substituted hydroxyapatite granules. *J. Mater. Sci.-Mater. Med.* **2002**, *13*, 1199–1206. [CrossRef]
4. Rawlings, R.D. Composition dependence of the bioactivity of glasses. *J. Mater. Sci. Lett.* **1992**, *11*, 1340–1343. [CrossRef]
5. Strnad, Z. Role of the glass phase in bioactive glass-ceramics. *Biomaterials* **1992**, *13*, 317–321. [CrossRef]
6. Shelby, J.E. *Introduction to Glass Science & Technology*; Royal Society of Chemistry: London, UK, 2020.
7. Kaimonov, M.R.; Safronova, T.V. Materials in the Na_2O–CaO–SiO_2–P_2O_5 System for Medical Applications. *Materials* **2023**, *16*, 5981. [CrossRef]
8. Filip, D.G.; Surdu, V.-A.; Paduraru, A.V.; Andronescu, E. Current Development in Biomaterials—Hydroxyapatite and Bioglass for Applications in Biomedical Field: A Review. *J. Funct. Biomater.* **2022**, *13*, 248. [CrossRef] [PubMed]
9. Arango-Ospina, M.; Xie, F.; Gonzalo Juan, I.; Riedel, R.; Ionescu, E.; Boccaccini, A.R. Review: Silicon oxycarbide based materials for biomedical applications. *Appl. Mater. Today* **2020**, *18*, 100482. [CrossRef]
10. Jones, J.R. Review of bioactive glass: From Hench to hybrids. *Acta Biomater.* **2013**, *9*, 4457–4486. [CrossRef] [PubMed]
11. Wu, X.; Yang, B.; Li, Y.; Yu, S. High temperature oxidation behaviors of C/C composites with SiOC coatings. *J. Eur. Ceram. Soc.* **2024**, *44*, 3537–3543. [CrossRef]
12. Widgeon, S.J.; Sen, S.; Mera, G.; Ionescu, E.; Riedel, R.; Navrotsky, A. ^{29}Si and ^{13}C Solid-State NMR Spectroscopic Study of Nanometer-Scale Structure and Mass Fractal Characteristics of Amorphous Polymer Derived Silicon Oxycarbide Ceramics. *Chem. Mater.* **2010**, *22*, 6221–6228. [CrossRef]
13. Wu, X.; Yu, S.; Li, Y. SiOC films on C/C composite prepared by chemical vapor deposition with hexamethyldisilazane precursor. *Ceram. Int.* **2022**, *48*, 20887–20894. [CrossRef]
14. Yan, J.; Wei, X.; Liu, Q.; Li, Y.; Yu, S. Microstructure and oxidation behavior of SiOC coatings on C/C composites co-deposited with HMDS and TEOS by using CVD process. *Ceram. Int.* **2024**, *50*, 7888–7896. [CrossRef]
15. Sen, S.; Widgeon, S.J.; Navrotsky, A.; Mera, G.; Tavakoli, A.; Ionescu, E.; Riedel, R. Carbon substitution for oxygen in silicates in planetary interiors. *Proc. Natl. Acad. Sci. USA* **2013**, *110*, 15904–15907. [CrossRef]
16. Gonzalo-Juan, I.; Detsch, R.; Mathur, S.; Ionescu, E.; Boccaccini, A.R.; Riedel, R. Synthesis and In Vitro Activity Assessment of Novel Silicon Oxycarbide-Based Bioactive Glasses. *Materials* **2016**, *9*, 959. [CrossRef] [PubMed]

17. Ren, N.; Li, J.; Qiu, J.; Sang, Y.; Jiang, H.; Boughton, R.I.; Huang, L.; Huang, W.; Liu, H. Nanostructured Titanate with Different Metal Ions on the Surface of Metallic Titanium: A Facile Approach for Regulation of rBMSCs Fate on Titanium Implants. *Small* **2014**, *10*, 3169–3180. [CrossRef] [PubMed]
18. Lepry, W.C.; Griffanti, G.; Nazhat, S.N. Bioactive sol-gel borate glasses with magnesium. *J. Non-Cryst. Solids* **2022**, *581*, 121415. [CrossRef]
19. Shearer, A.; Molinaro, M.; Montazerian, M.; Sly, J.J.; Miola, M.; Baino, F.; Mauro, J.C. The unexplored role of alkali and alkaline earth elements (ALAEs) on the structure, processing, and biological effects of bioactive glasses. *Biomater. Sci.* **2024**, *12*, 2521–2560. [CrossRef] [PubMed]
20. Xie, F.; Gonzalo-Juan, I.; Breitzke, H.; Fasel, C.; Trapp, M.; Buntkowsky, G.; Kleebe, H.-J.; Riedel, R.; Boccaccini, A.R.; Ionescu, E. Effect of Ca and B incorporation into silicon oxycarbide on its microstructure and phase composition. *J. Am. Ceram. Soc.* **2019**, *102*, 7645–7655. [CrossRef]
21. Chen, H.; Yu, S.; Li, Y. Bioactivity of SiOC and SiCaMgOC coatings prepared by sol-gel method. *Ceram. Int.* **2023**, *49*, 34248–34254. [CrossRef]
22. Ionescu, E.; Sen, S.; Mera, G.; Navrotsky, A. Structure, energetics and bioactivity of silicon oxycarbide-based amorphous ceramics with highly connected networks. *J. Eur. Ceram. Soc.* **2018**, *38*, 1311–1319. [CrossRef]
23. Vivanco-Benavides, L.E.; Martínez-González, C.L.; Mercado-Zúñiga, C.; Torres-Torres, C. Machine learning and materials informatics approaches in the analysis of physical properties of carbon nanotubes: A review. *Comput. Mater. Sci* **2022**, *201*, 110939. [CrossRef]
24. Kokubo, T.; Kushitani, H.; Sakka, S.; Kitsugi, T.; Yamamuro, T. Solutions able to reproduce in vivo surface-structure changes in bioactive glass-ceramic A-W. *J. Biomed. Mater. Res.* **1990**, *24*, 721–734. [CrossRef] [PubMed]
25. Goncalves, M.J.; Cruz, M.T.; Cavaleiro, C.; Lopes, M.C.; Salgueiro, L. Chemical, antifungal and cytotoxic evaluation of the essential oil of Thymus zygis subsp sylvestris. *Ind. Crop. Prod.* **2010**, *32*, 70–75. [CrossRef]
26. Koudehi, M.F.; Fooladi, A.A.I.; Mansoori, K.; Jamalpoor, Z.; Amiri, A.; Nourani, M.R. Preparation and evaluation of novel nano-bioglass/gelatin conduit for peripheral nerve regeneration. *J. Mater. Sci.-Mater. Med.* **2014**, *25*, 363–373. [CrossRef]
27. Wang, B.L.; Li, L.; Zheng, Y.F. In vitro cytotoxicity and hemocompatibility studies of Ti-Nb, Ti-Nb-Zr and Ti-Nb-Hf biomedical shape memory alloys. *Biomed. Mater.* **2010**, *5*, 044102. [CrossRef] [PubMed]
28. Jabir, M.S.; Sulaiman, G.M.; Taqi, Z.J.; Li, D. Iraqi propolis increases degradation of IL-1β and NLRC4 by autophagy following Pseudomonas aeruginosa infection. *Microb. Infect.* **2018**, *20*, 89–100. [CrossRef] [PubMed]
29. Zanoni, M.; Piccinini, F.; Arienti, C.; Zamagni, A.; Santi, S.; Polico, R.; Bevilacqua, A.; Tesei, A. 3D tumor spheroid models for in vitro therapeutic screening: A systematic approach to enhance the biological relevance of data obtained. *Sci. Rep.* **2016**, *6*, 19103. [CrossRef]
30. Payne, M.C.; Inkson, J.C. Longitudinal-optic-transverse-optic vibrational mode splittings in tetrahedral network glasses. *J. Non-Cryst. Solids* **1984**, *68*, 351–360. [CrossRef]
31. Lyu, Y.; Tang, H.; Zhao, G. Effect of Hf and B incorporation on the SiOC precursor architecture and high-temperature oxidation behavior of SiHfBOC ceramics. *J. Eur. Ceram. Soc.* **2020**, *40*, 324–332. [CrossRef]
32. Niemiec, W.; Szczygieł, P.; Jeleń, P.; Handke, M. IR investigation on silicon oxycarbide structure obtained from precursors with 1:1 silicon to carbon atoms ratio and various carbon atoms distribution. *J. Mol. Struct.* **2018**, *1164*, 217–226. [CrossRef]
33. Haider, A.; Haider, S.; Han, S.S.; Kang, I.-K. Recent advances in the synthesis, functionalization and biomedical applications of hydroxyapatite: A review. *RSC Adv.* **2017**, *7*, 7442–7458. [CrossRef]
34. Chen, H.; Wu, X.; Zhou, Y.; Yin, Y.; Yu, S.; Li, Y. Improved biological compatibility of carbon materials via aqua regia and CaCl2 treatments. *Mater. Chem. Phys.* **2023**, *296*, 127359. [CrossRef]
35. Sroka-Bartnicka, A.; Borkowski, L.; Ginalska, G.; Ślósarczyk, A.; Kazarian, S.G. Structural transformation of synthetic hydroxyapatite under simulated in vivo conditions studied with ATR-FTIR spectroscopic imaging. *Spectrochim. Acta Part A* **2017**, *171*, 155–161. [CrossRef]
36. Chen, J.; Zeng, L.; Chen, X.; Liao, T.; Zheng, J. Preparation and characterization of bioactive glass tablets and evaluation of bioactivity and cytotoxicity in vitro. *Bioact. Mater.* **2018**, *3*, 315–321. [CrossRef] [PubMed]
37. Vallet-Regi, M.; Salinas, A.J.; Román, J.; Gil, M. Effect of magnesium content on the in vitro bioactivity of CaO-MgO-SiO_2-P_2O_5 sol-gel glasses. *J. Mater. Chem.* **1999**, *9*, 515–518. [CrossRef]
38. Souza, M.T.; Crovace, M.C.; Schröder, C.; Eckert, H.; Peitl, O.; Zanotto, E.D. Effect of magnesium ion incorporation on the thermal stability, dissolution behavior and bioactivity in Bioglass-derived glasses. *J. Non-Cryst. Solids* **2013**, *382*, 57–65. [CrossRef]
39. Hench, L.L. Bioceramics: From Concept to Clinic. *J. Am. Ceram. Soc.* **1991**, *74*, 1487–1510. [CrossRef]
40. Maçon, A.L.B.; Kim, T.B.; Valliant, E.M.; Goetschius, K.; Brow, R.K.; Day, D.E.; Hoppe, A.; Boccaccini, A.R.; Kim, I.Y.; Ohtsuki, C.; et al. A unified in vitro evaluation for apatite-forming ability of bioactive glasses and their variants. *J. Mater. Sci. Mater. Med.* **2015**, *26*, 115. [CrossRef] [PubMed]
41. Cerruti, M.; Greenspan, D.; Powers, K. Effect of pH and ionic strength on the reactivity of Bioglass® 45S5. *Biomaterials* **2005**, *26*, 1665–1674. [CrossRef] [PubMed]

materials

MDPI

Article

The Impact of the Final Sintering Temperature on the Microstructure and Dielectric Properties of $Ba_{0.75}Ca_{0.25}TiO_3$ Perovskite Ceramics

Kamil Feliksik [1], Małgorzata Adamczyk-Habrajska [1], Jolanta Makowska [1,*], Joanna A. Bartkowska [1], Tomasz Pikula [2], Rafał Panek [3] and Oliwia Starczewska [1]

[1] Faculty of Science and Technology, Institute of Materials Engineering, University of Silesia, 75 Pułku Piechoty 1A, 41-500 Chorzow, Poland; kamil.feliksik@us.edu.pl (K.F.); malgorzata.adamczyk-habrajska@us.edu.pl (M.A.-H.); joanna.bartkowska@us.edu.pl (J.A.B.); oliwia.starczewska@us.edu.pl (O.S.)

[2] Institute of Electronics and Information Technology, University of Technology, 38A Nadbystrzycka Str., 20-618 Lublin, Poland; t.pikula@pollub.pl

[3] Department of Construction Materials Engineering and Geoengineering, Lublin University of Technology, 40 Nadbystrzycka Str., 20-618 Lublin, Poland; r.panek@pollub.pl

* Correspondence: jolanta.makowska@us.edu.pl

Abstract: $Ba_{0.75}Ca_{0.25}TiO_3$ ceramics were successfully synthesized by a simple solid-state reaction method. This study examined the influence of sintering temperature on the structure, microstructure, dielectric properties and electrical behavior of the material. The XRD analysis reveals that the tetragonal phase (P4mm) is dominant in all the synthesized materials, with those sintered at T = 1400 °C and T = 1450 °C being single-phase, while others exhibit a minor orthorhombic phase (Pbnm). Higher sintering temperatures promoted better grain boundary formation and larger grain sizes. The electric permittivity increased with temperature up to T = 1400 °C, followed by a sharp decline at T = 1450 °C. Additionally, the Curie temperature decreased with increasing sintering temperature, indicating changes in phase transition characteristics. Thermal analysis showed that higher sintering temperatures led to sharper heat capacity peaks, while pyroelectric and thermally stimulated depolarization currents were maximized at T = 1400 °C due to oxygen vacancies. These findings highlight the significant impact of sintering temperature on the material's structural and functional properties.

Keywords: $Ba_{0.75}Ca_{0.25}TiO_3$ ceramics; structure; microstructure; dielectric properties; thermal analysis; pyroelectric currents; perovskites

1. Introduction

Perovskites are a broad group of materials that are constantly attracting great interest due to their strong ferroelectric properties and piezoelectric and pyroelectric behavior and the possibility of influencing these properties by appropriately doping the base material and selecting appropriate technological process conditions. The perovskite family is described by a general formula of the form ABO_3, in which the A and B positions are usually occupied by metal ions. The flagship representatives of ferroelectric perovskites include barium titanate, $BaTiO_3$ (BT). Barium titanate is a ferroelectric compound that has a high dielectric constant and low dielectric loss. Therefore, it is one of the most widely used ceramic materials in electronic equipment [1,2]. The key applications of BT-type ceramics include positive-temperature-coefficient devices, pulse generators, infrared detectors, multilayer ceramic capacitors, voltage-tuning devices in microwave electronics, piezoelectric actuators, transducers, and charge storage devices [3–8]. These materials continue to attract significant interest not only for their properties but also because they are lead-free and environmentally friendly. This contrasts with popular $Pb(Zr,Ti)O_3$ ceramics, which are toxic due to their lead content [9–12].

Undoped barium titanate is historically the first lead-free ferroelectric and piezoelectric material. In order to improve its physical properties, attempts were made to modify it with ions of various elements. For example, in order to improve piezoelectric properties, grain size control was started by modifying BT with dopant ions such as Ca, Zr, Sn and Hf. Undoped BT ceramic material showed piezoelectric coefficient d33 values at the level of 338–519 pC/N. However, after modification with dopant ions, the piezoelectric coefficient d33 value increased to 625 pC/N [7,13,14].

Many studies have been carried out to improve the dielectric properties of barium titanate. These studies have involved substituting different ions at the A or B positions in the perovskite structure of BT ceramics [15]. Namely, for example, P Wang et al. doped BT ceramic material with calcium and zirconium ions. Calcium ions were substituted at the A position, while zirconium ions were substituted at the B position in the ABO_3 perovskite structure. The final ceramic material ($(Ba_{0.85}Ca_{0.15}(Ti_{0.9}Zr_{0.1})O_3$ (BCTZ)) was obtained by the conventional method of mixing oxides at different calcination and sintering temperatures. By selecting the dopant ions and selecting the appropriate technological conditions, the authors obtained a ceramic material that is a very promising lead-free material of importance for a multitude of applications. The authors showed that the BCTZ ceramics, which were calcined at $T = 1300$ °C and sintered at $T = 1540$ °C, exhibited optimal electrical properties, i.e., $d_{33} = 650$ pC/N, $d_{31} = 74$ pC/N, $k_p = 0.53$, $k_t = 0.38$, $k_{31} = 0.309$, $\varepsilon_r = 4500$ and $P_r = 11.69$ μC/cm^2 [16].

BT is characterized by its piezoelectric properties having a high dependence on temperature. In order to reduce this dependence and stabilize the piezoelectric properties, P. Zheng et al. attempted to dope $BaTiO_3$ with Zr ions, which were substituted for Ti ions. They obtained a ceramic material, $BaTi_{0.9625}Zr_{0.0375}O_3$, which was then modified with a small amount of CuO. The ceramic material obtained in this way has excellent piezoelectric properties, i.e., a piezoelectric coefficient of $d_{33} = 300$ pC/N, $k_p = 0.493$ and $k_{33} = 0.651$ with a tangent = 0.011, and a kp greater than 0.40 in a wide temperature range, namely from $T = 43$ °C to $T = 73$ °C, and its properties are almost constant in the range from $T = 25$ °C to $T = 55$ °C. The results from the cited authors indicate that the CuOBaTi, ZrO_3-modified ceramics obtained by them are a promising lead free material for practical applications, and appropriate doping significantly improves the properties of the tested material [17]. In turn, the work of A. Rached et al. presents the results of structural and electrical studies performed for barium titanate doped with erbium. The ceramic material $Ba_{0.91}Er_{0.1/3}Ca_{0.04}Ti_{0.92}Sn_{0.08}O_3$ was obtained by the solid-state diffusion method. The obtained material is characterized by a polycrystalline microstructure, as shown by SEM studies. Electrical conductivity studies allowed the determination of the activation energy, which is in the range from 1.3 eV to 0.213 eV and is dependent on temperature. Nyquist plots created for the obtained ceramics suggest that that the electrical properties of this compound are influenced by grain effects. The dielectric measurements performed indicate a high value of electric permittivity and reveal a phase transition from the ferroelectric to the paraelectric phase. Thanks to such doping, it was possible to increase the multifunctionality of barium titanate by improving its physical properties [18]. V. Paunovic et al. investigated the impact of Nb^{5+} on the microstructure and dielectric properties of $BaTiO_3$ ceramics, using the solid-state method to produce Nb/Mn–$BaTiO_3$. They found that increasing the Nb^{5+} content reduces the electric permittivity, with the highest values observed for 0.1 Nb/Mn–$BaTiO_3$ sintered at $T = 1350$ °C. A higher Nb content led to lower electric permittivity and loss tangent values, as well as a decrease in the Curie constant. The optimal dielectric properties were observed in ceramics doped with 0.1 at% Nb, and the effect of Nb^{5+} depended on the sintering temperature [19].

Researchers have doped barium titanate with rare earth ions like La^{3+}, Gd_{3+}, Dy^{3+} and Yb^{3+} to improve its electrical properties. Doping with Sr^{2+} and Ca^{2+} increases piezoelectricity and electrical breakdown strength. On the other hand, doping with rare earth ions lowers the phase transition temperature and affects electrical properties. Lanthanum and gadolinium improve electrical performance, while dysprosium and ytterbium worsen it [20–27]. $BaTiO_3$ is an excellent ferroelectric material, but by appropriate doping with iron ions it can be enriched with magnetic properties. The work of S.B. Bhoobas et al. presents interesting results from research on iron-modified barium titanate, i.e., $FeBaTiO_3$. The authors report that barium titanate doped in this way retains its crystallinity in the tetragonal phase with the P4mm space group, with only a small change in the tetragonality ratio, the value of which changes from 1.008 ($BaTiO_3$) to 1.002 ($FeBaTiO_3$). The ferroelectric–paraelectric phase transition temperature for the modified $FeBaTiO_3$ was shifted from 120 °C (for undoped $BaTiO_3$) to 300 °C (for modified $FeBaTiO_3$). Such properties suggest that this material can be used for high-temperature capacitive applications. Magnetic hysteresis (M-H) studies at room temperature performed on modified $FeBaTiO_3$ show ferromagnetic properties [28–30].

Among so many modifiers, it is difficult to choose the most optimal one. An analysis of the literature data has shown that the most comprehensive benefits can be achieved by doping barium titanate with calcium ions. Primarily, by improving sinterability, they influence the stabilization of the crystal structure [31,32]. Additionally, Ca^{2+} ion doping leads to a significant reduction in dielectric losses while simultaneously increasing electric permittivity, which is advantageous for applications such as ceramic capacitors [33]. In these applications, the improvement in dielectric breakdown strength is also important, which is ensured by Ca^{2+} doping. Moreover, doping with calcium ions enhances piezoelectric properties [34], which significantly expands the application potential of $BaTiO_3$-based materials.

Literature reports show how much influence the selection of the appropriate type and amount of $BaTiO_3$-modifying ions has on the piezoelectric and electrical properties and their stability. The selection of appropriate technological conditions, including sintering temperature, also has a significant influence. That is, electrical properties, similarly to piezoelectric properties, can be controlled by introducing appropriate doping ions into the BT base material and strictly controlling both the chemical composition and conditions during the technological process [35–44].

There are several methods for obtaining BT ceramic material, both undoped and doped. These include the sol–gel method [1,45–48], spark plasma sintering [49,50] and solid-state synthesis. Due to the fact that the ceramic material presented in this paper may have great application significance, the authors used the solid-state synthesis method to produce it. This method, unlike other methods, allows for the production of ceramic material on an industrial scale and is an economical method.

Based on the analysis of literature data and the analysis of factors influencing the electrical properties of the modified BT ceramics, an attempt was made to dope $BaTiO_3$ with calcium ions and investigate the effect of the sintering temperature on the properties of these modified ceramics. The aim of this work was to obtain $Ba_{0.75}Ca_{0.25}TiO_3$ ceramic material and to investigate the effect of sintering temperature on the structural, microstructural, and dielectric properties. The tested $Ba_{0.75}Ca_{0.25}TiO_3$ ceramic material was prepared using a mixed oxide and carbon method and synthesized by the free sintering method.

2. Materials and Methods

$Ba_{0.75}Ca_{0.25}TiO_3$ ceramics were synthesized using barium carbonate ($BaCO_3$, Avantor (Radnor Township, PA, USA, 99.99%), calcium carbonate ($CaCO_3$, Avantor (Radnor Township, PA, USA), 99.99%) and titanium oxide (TiO_2, Avantor (Radnor Township, PA, USA, 99.99%) as starting materials. The oxide and carbonate powders were first weighed in stoichiometric proportions, then mixed in a mortar before being further processed in a planetary ball mill. The wet milling process used ethanol (Avantor Radnor Township, PA, USA, 99.99%) as a solvent and zirconium–yttrium grinding media (diameter d = 1 cm) in polyamide containers for a duration of t = 24 h. After milling, the mixture was air-dried, followed by calcination of the dried powders in a muffle furnace at T = 1000 °C for t = 3 h.

During synthesis, the furnace heating rate was set to $v = 5$ °C/min. The programmable temperature control ensured a linear temperature increase and stable temperature maintenance. The cooling rate of the synthesized samples corresponded to the free cooling rate of the furnace.

The synthesized powders were first mixed in a mortar, followed by wet milling in a planetary ball mill using ethanol as the milling medium. Zirconium–yttrium grinding media with a diameter of 1 cm were employed, and the process was conducted in polyamide containers for a duration of $t = 24$ h. After milling, the powders were air-dried. The dried powders were then pressed into compacts with a diameter of 10 mm using a semi-automatic hydraulic press at a pressure of $p = 300$ MPa in steel dies. These compacts were subsequently placed in corundum crucibles on a bed of alumina (Al_2O_3, Avantor, 99.99%) for further processing. The compacts were sintered in an electric resistance furnace at various temperatures—$T_S = 1250$ °C, $T_S = 1300$ °C, $T_S = 1350$ °C, $T_S = 1400$ °C and $T_S = 1450$ °C—for $t = 3$ h, in an air atmosphere. For the sintering process, the heating rate was the same as during synthesis. The samples for dielectric measurements were prepared in the form of disks with a thickness of $d = 0.6$ mm and a surface area of 1 cm^2, with parallel surfaces coated with conductive silver paste (P-120, supplied by the Polish State Mint, Warsaw, Poland) using the burn-in method.

The phase composition of the prepared samples was evaluated using X-ray diffraction (XRD) on a Panalytical X'pert PRO MPD diffractometer (Eindhoven, The Netherlands) in the standard θ–2θ configuration. The diffractometer was equipped with a PW 3050 goniometer, a PIXcel 1D detector, and a copper (Cu) radiation source (CuKα = 1.54178 Å). For phase and structural analysis, the HighScore Plus 3.0e software was used, incorporating the 2022 edition of the PDF-2 DL database from JCPDS-ICDD.

The Archimedes method with distilled water was employed to evaluate the sample density. The microstructure of the produced ceramic samples, along with qualitative and quantitative EDS analysis, was examined using a JEOL JSM-7100F TTL LV scanning electron microscope (Akishima, Japan) equipped with an EDS system.

The dielectric properties were measured on specially prepared samples using a computerized system that included an Agilent E4980A LCR meter (Santa Clara, CA, USA). To reduce stresses from prior mechanical processing and allow for the relaxation of immobile defects formed during sintering, the samples were pre-heated at $T = 100$ °C for $t = 30$ min prior to measurements (heating rate $v = 5$ °C/min).

The enthalpy and characteristic temperatures of the ceramic samples were determined using differential scanning calorimetry (DSC1, Mettler Toledo, Columbus, OH, USA). BCT25 ceramic powders were weighed with an analytical balance (RADWAG AS 60/220/C/2) and encapsulated in 40 μL hermetically sealed aluminum crucibles. Measurements were conducted from room temperature to $T = 300°$ C at a heating rate of $T = 10$ °C/min, using nitrogen as the drying gas. Results were analyzed with STARe software (Version 11.00a).

The temperature dependencies of pyroelectric currents and thermally stimulated depolarization (TSD) currents were measured using a computerized system with a Keithley 6485 picoammeter (Cleveland, OH, USA). Before the measurement, the samples were polarized in a DC field at $T = 150$ °C for 20 min and then cooled to room temperature in this field. The samples were then heated at a rate of 5 K/min up to 773 K, and the current was measured as a function of temperature and time.

3. Results and Discussion

3.1. X-Ray Analysis

Figure 1 shows the XRD patterns registered for all the studied $Ba_{0.75}Ca_{0.25}TiO_3$ samples. It can be noted that the tetragonal phase (space group P4mm), characteristic of $BaTiO_3$ at room temperature, dominates.

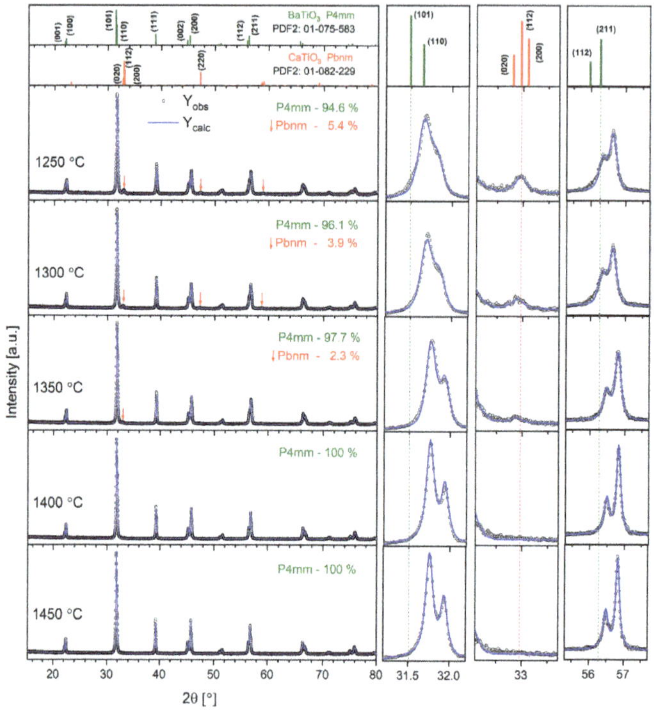

Figure 1. The results of XRD investigations for all the BCT samples studied. The right panel shows the magnification of the diffractograms in the 2θ ranges of 31–32°, 32–32.5° and 55–58°. The top panel shows standard patterns for tetragonal $BaTiO_3$ (PDF2 card number 01-075-583) and orthorhombic $CaTiO_3$ (PDF2 card number 01-082-229) as a reference.

Ceramics sintered at T_S = 1400 °C and T_S = 1450 °C are single-phase, while a small amount of orthorhombic-phase ceramics (Pbnm space group), characteristic of $CaTiO_3$ perovskite, was detected for the other samples. The top panel of Figure 1 shows standard XRD patterns for the tetragonal $BaTiO_3$ and orthorhombic $CaTiO_3$ as a reference. Thus, it can be assumed that only the highest temperatures, T = 1400 °C and T = 1450 °C, are sufficient to dissolve all the Ca^{2+} ions in the BTO lattice. On the other hand, a synthesis temperature of up to T = 1350 °C provides too little energy, and the Ca-rich orthorhombic phase is separated. Moreover, the contribution of the Pbnm phase gradually decreases with the growth of the sintering temperature, confirming the above findings. In turn, the volume of the Pbnm unit cell significantly increases with the rise in synthesis temperature, proving that some amount of Ba^{2+} ions replace Ca^{2+} ions in the orthorhombic Pbnm structure. To obtain the structural parameters, the Rietveld refinement method was applied. The results are displayed in Table 1. One can note the gradual decrease in lattice parameters with the increase in annealing temperature for the P4mm majority phase. This is mostly because more Ca^{2+} ions are located inside the BTO structure when the annealing temperature increases. The latter can be evidenced by the gradual shifting of diffraction peaks towards the higher 2θ angle visible in the right panel of Figure 1. For ceramics sintered at T_S = 1400 °C and T_S = 1450 °C, the lattice parameters are almost identical, as all the Ca has been dissolved in the BTO lattice. It is clearly visible in the right panel of Figure 1 that the width of the diffraction peaks gradually decreases with the increase in annealing temperature. This results from the lower level of lattice strains and imperfections caused by the elevated temperature.

Table 1. Structural parameters derived from Rietveld refinement procedure.

T_S [°C]	Crystal System	Space Group	Lattice Parameters [A]	Volume [A^3]	Contrib. [wt. %]
1250	tetragonal	P4mm	a = b = 3.9679 c = 4.0068	63.08	94.6
	orthorhom.	Pbnm	a = 5.4548 b = 7.6615 c = 5.4083	226.02	5.4
1300	tetragonal	P4mm	a = b = 3.9646 c = 4.0032	62.92	96.1
	orthorhom.	Pbnm	a = 5.4628 b = 7.6767 c = 5.4163	227.14	3.9
1350	tetragonal	P4mm	a = b = 3.9599 c = 4.0012	62.74	97.7
	orthorhom.	Pbnm	a = 5.4945 b = 7.6855 c = 5.4208	228.9	2.3
1400	tetragonal	P4mm	a = b = 3.9565 c = 3.9999	62.62	100
1450	tetragonal	P4mm	a = b = 3.9564 c = 3.9997	62.60	100

3.2. Density and Microstructure

In the first stage of the research, the real density of the samples was determined using the Archimedes method (Table 2). The density of the sample sintered at the lowest temperature had the smallest value of $\rho = 4.47\ g/cm^3$, while for the material sintered at T_S = 1400 °C, this value was the highest, at $\rho = 5.67\ g/cm^3$. It is also worth noting that the density of the ceramic material sintered at T_S = 1450 °C was slightly lower, at $\rho = 5.47\ g/cm^3$.

Table 2. Density and average grain size of $Ba_{0.75}Ca_{0.25}TiO_3$ ceramics sintered at different temperatures.

T_S [°C]	Average Grain Size [μm]	Experimental Density [g/cm³]
1250	-	4.47
1300	-	5.11
1350	0.70	5.41
1400	1.60	5.67
1450	5.4	5.47

The results presented in Table 2 clearly indicate that as the final sintering temperature increases, the material's density significantly increases, which leads to a reduction in its porosity through diffusion kinetics [51]. The pores located at the grain boundaries significantly hinder their movement, which slows down grain growth. Once they are eliminated, grain growth becomes faster [1,41–52]. Therefore, it can be expected that increasing the final sintering temperature will lead to an increase in grain size. This theory was confirmed by microstructural studies conducted using a scanning electron microscope. Microstructure images of the ceramic materials discussed are shown in Figure 2. Similar results were observed in BZT-BCT 50/50 ceramics [53], where the authors explained the increase in grain size and reduction in pore size with rising sintering temperatures using the phenomenological kinetic grain growth equation, which suggests that grain size increases as sintering temperature rises [54].

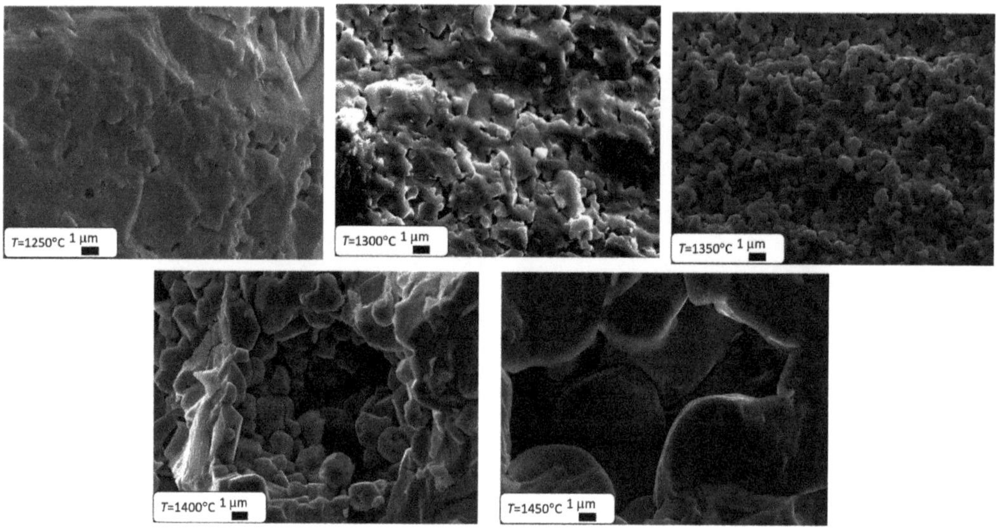

Figure 2. SEM images of $Ba_{0.75}Ca_{0.25}TiO_3$ ceramics sintered at different temperatures.

In the case of the two lowest temperatures, the grain structure did not form. The grain structure only develops in ceramics sintered at T_S = 1350 °C. Further increases in sintering temperature lead to a better formation of grain boundaries and a significant increase in grain size (Table 2). A similar grain growth caused by an increase in sintering temperature is observed in $Ba_{0.70}Ca_{0.30}TiO_3$ ceramics [55], as well as in many other materials based on barium titanate, such as $BaZr_{0.2}Ti_{0.8}O_3$ [56] or BZT-BCT 50-50 [53]. The main factor behind grain growth is the decrease in the total grain boundary area, which lowers the interfacial free energy of the material [57]. Since interfacial energy is inversely related to grain size [58], this leads to a reduction in the overall grain boundary area. As the grains grow larger, the porosity of the structure decreases [59].

3.3. Dielectric Properties

In order to investigate the effect of the final sintering temperature on the dielectric parameters of $Ba_{0.75}Ca_{0.25}TiO_3$ ceramics, measurements of the temperature changes in electric permittivity (ε') and the loss tangent (tan δ) were carried out for all five samples. Figure 3 shows such dependences obtained at a frequency of the measuring field equal to 1 kHz.

The maximum value of electric permittivity (ε'_{max}) gradually increases with the rise in sintering temperature (Table 3). This trend persists until T_S = 1400 °C. At T_S = 1450 °C, a sharp decrease in ε'_{max} is observed. Such an effect has been observed, among others, for $BaTiO_3$ ceramics [60] and PBZT 37/70/30 ceramics [61]. This phenomenon can be explained based on the defect model [62], which assumes that the factor suppressing the formation of the domain structure is crystal lattice defects, which tend to localize at grain boundaries. As the grain size increases, the contribution of grain boundaries to the sample volume decreases, leading to a reduction in the number of defects—allowing the domain structure to grow more freely [62]. Moreover, the increase in grain size leads to a decrease in the thickness of the relatively more insulating grain boundary layer [63]. As a result of these processes, an increase in electric permittivity is observed. Therefore, electric permittivity increases with the increase in grain size. The described trend breaks down for BCT ceramics sintered at 1450 °C, where the electric permittivity decreases across the entire measured temperature range. This is due to significant and rapid grain growth; the average grain size for this material is 5.6 μm, which is over three times larger than the average grain size of ceramics sintered at 1400 °C (1.6 μm, see Table 2). It is well known that in ceramics made from barium titanate, dielectric properties, particularly the value of the electric permittivity, strongly depend on grain size, and this relationship is not straightforward. Specifically, the authors of studies [60,64] found that in $BaTiO_3$ ceramics, the optimal grain size for achieving the highest dielectric constant is 0.8 μm. Further reduction in grain size leads to a decrease in electric permittivity, which can be explained by the increased contribution of non-ferroelectric grain boundaries to the overall volume of the sample [65,66]. Increasing the grain size also leads to a decrease in the dielectric constant. In $BaTiO_3$ ceramics with grain sizes in the range of 0.8–50 μm, there is an inversely proportional relationship between grain size and the dielectric constant. According to Buessem [67], this phenomenon is associated with an increase in residual stress within individual grains, accompanied by an increase in tetragonal distortion. It is also worth mentioning that in other systems based on barium titanate, the influence of sintering temperature on the increase in the maximum value of electric permittivity is observed. Examples of such systems include BZT-BC 50/50 [53] ceramics and $(Ba_{0.85}Ca_{0.15})(Sn_xZr_{0.1x}Ti_{0.90})O_3$ ceramics [68]. In the latter case, the maximum electric permittivity (ε_{max}) reaches its peak value at a sintering temperature of T_S = 1450 °C.

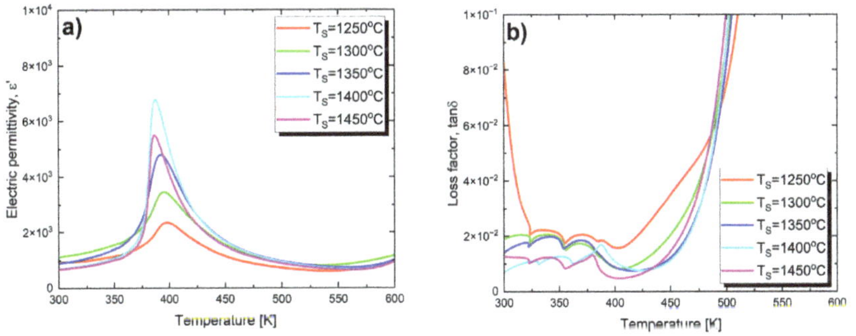

Figure 3. Electric permittivity (**a**) and loss tangent (**b**) as a function of temperature, measured at a frequency of 1 kHz, on heating processes for $Ba_{0.75}Ca_{0.25}TiO_3$ ceramics sintered at various temperatures T_S.

Table 3. The influence of the sintering temperature (T_S) on the dielectric parameters of $Ba_{0.75}Ca_{0.25}TiO_3$ ceramics measured at a frequency of f = 1 kHz.

T_S [°C]	T_C [K]	ε_{RT}	ε_{max}	$tg\delta_{RT}$	$tg\delta_{TC}$	T_0 [K]	C [K]	T_{dev} [K]	γ
1250	397.49	922.4	2345.712	0.119	0.016	373	5.9×10^4	421	1.85
1300	395.01	1075.9	3450.791	0.016	0.009	376	1.2×10^5	418	1.71
1350	392.09	839.9	4799.394	0.012	0.011	380	1.5×10^5	415	1.60
1400	387.40	657.9	6772.963	0.007	0.014	364	1.1×10^5	406	1.31
1450	385.39	660.6	5492.813	0.013	0.008	375	1.1×10^5	407	1.26

In the context of decreasing dielectric permittivity, it is also important to note the reduction in the porosity of materials. As the temperature increases, the porosity of the materials changes from 16% for ceramics sintered at T_S = 1250 °C to 1.5% for ceramics sintered at 1450 °C. Such a significant change in porosity affects the value of dielectric permittivity, potentially increasing it by a significant factor, as demonstrated by the authors of the study [69]. The changing grain/particle sizes and the proportion of pores in the total volume of the sample shape the properties not only of electroceramic materials but also other materials. Examples include graphene substrates [70] or thermoelectric materials [71].

Changes in the final sintering temperature also affect the value of the maximum electric permittivity temperature, which can be associated with the phase transition temperature between the low-temperature ferroelectric phase and the paraelectric phase. As T_S increases, the Curie temperature (T_C) decreases (Table 3).

Additionally, the maxima of dielectric permittivity observed in Figure 3 exhibit broadening, which appears to be more pronounced at lower sintering temperatures. To fully assess this phase transition broadening, an attempt was made to fit both the classical and modified Curie–Weiss laws to the experimental data. Since the $1/\varepsilon(T)$ dependence does not immediately become linear above the T_C temperature, the classical Curie–Weiss law is only applied above the T_{dev} temperature. In the temperature range between T_C and T_{dev}, the modified Curie–Weiss law is used (1):

$$\frac{1}{\varepsilon'} - \frac{1}{\varepsilon'_{max}} = \frac{(T - T_C)^{\gamma}}{\delta} \tag{1}$$

where

ε_{max}—maximum value of electric permittivity;

δ—the Curie-like constant;

γ—the diffuseness parameter.

A plot of $\ln(1/\varepsilon - 1/\varepsilon_{max})$ versus $\ln(T - T_C)$ for the selected temperature (T_S = 1350 °C) is shown as an example in Figure 4. The obtained values of T_{dev}, T_0 and C are collected in Table 3. The value of the diffuseness parameter decreases significantly, indicating a reduction in the degree of phase transition diffuseness and an increase in the order of the crystal structure. The results obtained are in good correlation with SEM and XRD studies. The reduction in phase transition broadening due to an increase in sintering temperature was also observed in BZT ceramics [72]. In contrast, in BSN ceramics, raising the sintering temperature significantly increased the degree of phase transition broadening [68].

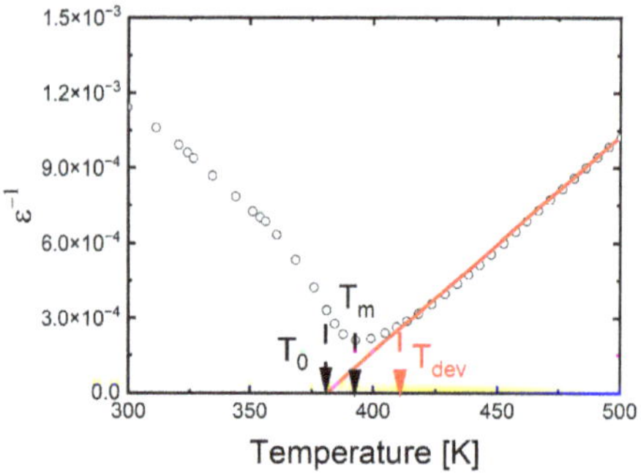

Figure 4. The reciprocal permittivity ($1/\varepsilon$) at 1 kHz as a function of temperature for $Ba_{0.75}Ca_{0.25}TiO_3$ ceramics sintered at various temperatures, T_S. Bullets indicate experimental data, red line is the fitting line.

The effect of the temperature changes on the loss factor is shown in Figure 3b. In the case of materials sintered at temperatures of T = 1250 °C, T = 1300 °C and T = 1350 °C, only slight changes in the values of the loss tangent (tan δ) are noted in the temperature range of 325 K–400 K. A distinct loss peak, observed at the Curie temperature and indicating a phase transition, only appears for ceramics sintered at the two highest temperatures. All the presented characteristics exhibit a strong increase in losses above 450 K, which may be attributed to the increased mobility of charge carriers arising from defects or vacancies in the sample [51,73]. The value of the loss tangent at room temperature significantly decreases with the increase in sintering temperature (Table 3), which is extremely important from an application standpoint.

3.4. Thermal Analysis

As mentioned above, the temperature characteristics of dielectric permittivity show a peak associated with the transition from the low-temperature ferroelectric phase to the paraelectric phase. To investigate the effect of sintering temperature on the thermal parameters of this transition, the temperature dependence of specific heat was determined through calorimetric measurements. Figure 5 shows examples of the temperature dependencies of the specific heat capacity (C_p) for $Ba_{0.75}Ca_{0.25}TiO_3$ sintered at two extreme values of T_S. The investigations were carried out during heating cycles.

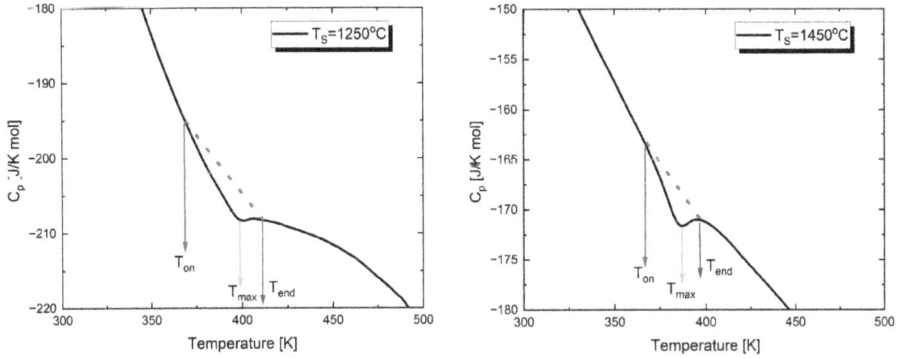

Figure 5. Temperature dependences of the specific heat capacity, C_p, measured upon heating of $Ba_{0.75}Ca_{0.25}TiO_3$ ceramics sintered at different temperatures.

In the presented characteristics, anomalies appear at temperatures (T_{max}) corresponding to those of the dielectric permittivity peaks. These values are slightly lower than the Curie temperature (T_C), which is a well-known behavior (Table 1) [74,75]. The temperatures corresponding to the onset (T_{on}) and completion (T_{end}) of the phase transition are also provided in Table 4. Additionally, Table 4 presents the values of the total enthalpy change, estimated from the area under the anomaly curve, using the background lines shown as dashed lines in Figure 5.

Table 4. The influence of sintering temperature (T_S) on the parameters of the FE PE phase transition in $Ba_{0.75}Ca_{0.25}TiO_3$ ceramics.

T_S [°C]	ΔH [J/g]	T_{on} [K]	T_{end} [K]	T_{max} [K]
1250	0.42	362.5	412.0	395.8
1300	0.37	368.9	414.8	392.0
1350	0.27	371.7	412.1	388.2
1400	0.26	362.5	404.7	384.3
1450	0.32	369.7	397.5	383.6

Analyzing the results presented in Table 4, it can be said that with the increase in sintering temperature, and consequently the development of a grain structure, the peak observed in the $C_p(T)$ dependencies becomes sharper. A similar trend was observed by the authors of study [76]; an increase in grain size leads to the sharpening of the anomaly. Additionally, it is worth noting that as the sintering temperature increases, the enthalpy decreases. These facts can be explained by the changing concentration of defects.

Specifically, a high concentration of defects [77] and the deformation field caused by dislocations [78] can significantly influence the thermodynamic properties of the material near the phase transition. The high density of defects leads to an increase in heat capacity and a broadening of the heat capacity jump [79].

3.5. Pyroelectric and Thermally Stimulated Depolarization Currents

To gain a deeper understanding of the dielectric properties and assess the potential applications of the studied materials, pyroelectric and thermally stimulated depolarization (TSD) current measurements were performed. The resulting characteristics are displayed in Figure 6.

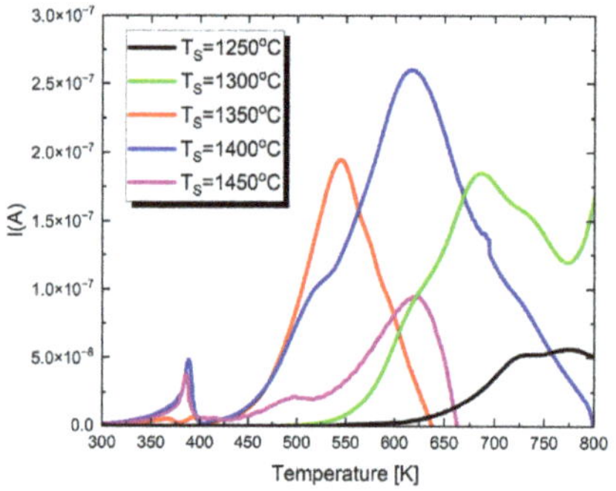

Figure 6. The temperature dependences of pyroelectric and thermally stimulated depolarization currents for $Ba_{0.75}Ca_{0.25}TiO_3$ ceramics sintered at different temperatures.

In the presented I(T) characteristics, depending on the sintering temperature, one or two maxima appear. The first low-temperature maximum appears only in the case of materials sintered at T = 1300 °C, T = 1350 °C and T = 1400 °C. Its occurrence temperature coincides, within the measurement error, with the TC temperature. Therefore, this is the maximum of the pyroelectric current. The next maximum, visible for all discussed ceramic materials, is associated with the presence of strong thermally stimulated depolarization currents in the material. These currents reach their highest value for ceramics sintered at T = 1400 °C. Thermally stimulated depolarization currents are connected with the presence of material oxygen vacancies. The proposed mechanism for generating TSDCs in $BaTiO_3$-based materials can be explained as follows [80]. During DC polarization at high temperatures, oxygen vacancies accumulate near the cathode and are depleted near the anode due to the blocking of ionic species by the metal electrodes. This results in a significant concentration gradient of $V_O^{\bullet\bullet}$. The spatial variation in the $V_O^{\bullet\bullet}$ concentration also induces gradients in electron and hole concentrations. After rapid cooling to lower temperatures, the DC field is removed, and the samples are short-circuited, leading to a metastable condition. The driving force for the depolarization of each species is the chemical potential created by these concentration gradients. Since electrons and holes move much faster than $V_O^{\bullet\bullet}$, their depolarization—producing an electronic current—occurs immediately after the samples are short-circuited. However, the polarized profile of $V_O^{\bullet\bullet}$ persists due to its much lower mobility. As the temperature increases, thermally activated depolarization of $V_O^{\bullet\bullet}$ takes place. This change in ionic polarization within the material generates image charges at the metal electrodes, producing the depolarization current. Therefore, the measured TSDC is governed by the relaxation of $V_O^{\bullet\bullet}$.

4. Conclusions

$Ba_{0.75}Ca_{0.25}TiO_3$ ceramics were obtained by the solid-state synthesis method from simple oxides and carbonates. XRD studies confirmed the partial substitution of Ba^{2+} ions with Ca^{2+} ions in the tetragonal P4mm lattice characteristic of a pure $BaTiO_3$ compound.

However, only the ceramics sintered at T_S = 1400 °C and T_S = 1450 °C turned out to be single-phase. The sintering of samples at a temperature below T = 1400 °C caused a separation of small amounts of the Ca-rich orthorhombic phase. Microstructural analysis confirmed that higher sintering temperatures promote better grain boundary formation and a substantial increase in grain size. As the sintering temperature increases, the maximum electric permittivity (ε'_{max}) rises up to T = 1400 °C but drops sharply at T = 1450 °C, which can be attributed to changes in the defect structure and domain formation. Simultaneously, an increase in the sintering temperature results in a reduction in the Curie temperature (T_C), indicating changes in the phase transition behavior between the ferroelectric and paraelectric phases. Thermal analysis shows that higher sintering temperatures lead to a sharper peak in heat capacity, which is related to the development of the grain structure and a reduction in the number of defects. Additionally, pyroelectric and thermally stimulated depolarization (TSD) currents reach their highest values at T = 1400 °C due to the presence of oxygen vacancies in the material.

The lack of lead makes $Ba_{0.75}Ca_{0.25}TiO_3$ ceramics a strong contender in fields where environmental concerns are paramount. The results presented in this paper serve as a starting point for further research aimed at improving the performance of this material, as well as enhancing its dielectric and piezoelectric properties and refining stability under various environmental conditions.

Author Contributions: Conceptualization, K.F. and M.A.-H.; methodology K.F. and M.A.-H.; SEM measurements and analysis—J.M.; XRD measurements and analysis, R.P. and T.P.; formal analysis, K.F. and M.A.-H.; investigation, SEM, dielectric properties, pyroelectric and thermally stimulated depolarization currents, DSCs, K.F.; DSC measurements, help in analyzing results—O.S.; writing—original draft preparation, K.F.; writing—review and editing, J.A.B. and M.A.-H.; visualization, J.A.B.; supervision, M.A.-H. All authors have read and agreed to the published version of the manuscript.

Funding: This project was co-financed by the European Union within the program "The European Funds for Śląsk (Silesia) 2021–2027".

Institutional Review Board Statement: Not applicable.

Informed Consent Statement: Not applicable.

Data Availability Statement: The original contributions presented in the study are included in the article, further inquiries can be directed to the corresponding author.

Conflicts of Interest: The authors declare no conflicts of interest.

References

1. Khedhri, M.H.; Abdelmoula, N.; Khemakhem, H.; Douali, R.; Dubois, F. Structural, spectroscopic and dielectric properties of Ca-doped $BaTiO_3$. *Appl. Phys. A* **2019**, *125*, 193. [CrossRef]
2. Gromov, O.G.; Kuz'min, A.P.; Kunshina, G.B.; Lokshin, E.P.; Kalinnikov, V.T. Preparation of Powder Barium Titanate. *Inorg. Mater.* **2006**, *42*, 176–181. [CrossRef]
3. Puli, V.S.; Martınez, V.R.; Kumar, A.; JScott, F.; Katiyar, R.S. A quaternary lead based perovskite structured materials with diffuse phase transition behavior. *Mater. Res. Bull.* **2011**, *46*, 2527–2530. [CrossRef]
4. Puli, V.S.; Kumar, A.; Chrisey, D.B.; Tomozawa, M.; Scott, J.F.; Katiyar, R.S. Barium zirconate-titanate/barium calcium-titanate ceramics via sol-gel process: Novel high-energy-density capacitors. *J. Phys. D Appl. Phys.* **2011**, *44*, 395403. [CrossRef]
5. Sharma, P.; Kumar, P.; Juneja, J.K.; Ahlawat, N.; Punia, R. Structural and dielectric properties of substituted barium titanate ceramics for capacitor applications. *Ceram. Int.* **2015**, *41 Part A*, 13425–13432. [CrossRef]
6. Chen, L.; Fan, H.; Li, Q. Characterization of acceptor-doped (Ba, Ca)TiO_3 "hard" piezoelectric ceramics for high-power applications. *Ceram. Int.* **2017**, *43*, 5579–5584. [CrossRef]
7. Chen, L.; Fan, H.; Zhang, S. Investigation of MnO_2-doped (Ba, Ca)TiO_3 lead-free ceramics for high power piezoelectric applications. *J. Am. Ceram. Soc.* **2017**, *100*, 3568–3576. [CrossRef]
8. Borkar, S.N.; Aggarwal, P.; Deshpande, V.K. Effect of calcium substitution on structural, dielectric, ferroelectric, piezoelectric, and energy storage properties of $BaTiO_3$. *Curr. Appl. Phys.* **2022**, *39*, 205–213. [CrossRef]
9. Aghayan, M.; Zak, A.K.; Behdani, M.; Hashim, A.M. Sol-gel combustion synthesis of Zr-doped $BaTiO_3$ nanopowders and ceramics: Dielectric and ferroelectric studies. *Ceram. Int.* **2014**, *40 Part B*, 16141–16146. [CrossRef]

10. Kaushal, A.; Olhero, S.M.; Singh, B.; Fagg, D.P.; Bdikin, I.; Ferreira, J.M.F. Impedance analysis of $0.5Ba(Zr_{0.2}Ti_{0.8})O_3$-$0.5(Ba_{0.7}Ca_{0.3})TiO_3$ ceramics consolidated from micro-granules. *Ceram. Int.* **2014**, *40 Part B*, 10593–10600. [CrossRef]
11. Jain, A.; Saroha, R.; Pastor, M.; Jha, A.K.; Panwar, A.K. Effect of sintering duration on structural and electrical properties of $Ba_{0.9}Sr_{0.1}Ti_{0.96}Zr_{0.04}O_3$ solid solution. *Curr. Appl. Phys.* **2016**, *16*, 859–866. [CrossRef]
12. Aneesh Kumar, K.S.; Bhowmik, R.N. Micro-structural characterization and magnetic study of $Ni_{1.5}Fe_{1.5}O_4$ ferrite synthesized through coprecipitation route at different pH values. *Mater. Chem. Phys.* **2014**, *146*, 159–169. [CrossRef]
13. Acosta, M.; Novak, N.; Jo, W.; Rodel, J. Relationship between the electromechanical properties and phase diagram in the Ba $(Zr_{0.2}Ti_{0.8})O_{3-x}(Ba_{0.7}Ca_{0.3})TiO_3$ lead-free piezoceramic. *Acta Mater.* **2014**, *80*, 48–55. [CrossRef]
14. Liu, X.; Tan, X. Crystal structure and electrical properties of lead-free (1-x)$BaTiO_{3-x}(Bi_{1/2}\ A_{1/2})TiO_3$ (A = Ag, Li, Na, K, Rb, Cs) ceramics. *J. Am. Ceram. Soc.* **2013**, *96*, 3425–3429. [CrossRef]
15. Ma, N.; Zhang, B.P.; Yang, W.G.; Guo, D. Phase structure and nano-domain in high performance of $aTiO_3$ piezoelectric ceramics. *J. Eur. Ceram. Soc.* **2012**, *32*, 1059. [CrossRef]
16. Wang, P.; Li, Y.; Lu, Y. Enhanced piezoelectric properties of $(Ba_{0.85}Ca_{0.15})$ $(Ti_{0.9}Zr_{0.1})$ O_3 lead-free ceramics by optimizing calcination and sintering temperature. *J. Eur. Ceram. Soc.* **2005**, *31*, 2005–2012. [CrossRef]
17. Zheng, P.; Zhang, J.L.; Shao, S.F.; Tan, Y.Q.; Wang, C.L. Piezoelectric properties and stabilities of CuO-modified $Ba(Ti,Zr)O_3$ ceramics. *Appl. Phys. Lett.* **2009**, *94*, 032902. [CrossRef]
18. Rached, A.; Wedern, M.A.; Belkahla, A.; Dhahri, J.; Khirouni, K.; Alaya, S.; Martín-Palma, R.J. Effect of doping in the physico-chemical properties of $BaTiO_3$ ceramics. *Physica B* **2020**, *596*, 412343. [CrossRef]
19. Paunovic, V.; Mitic, V.V.; Djordjevic, M.; Prijic, Z. Niobium doping effect on BaTiO3 structure and dielectric properties. *Ceram. Int.* **2020**, *46*, 8154–8164. [CrossRef]
20. Jain, A.; Panwar, A.K. Synergetic effect of rare-earths doping on the microstructural and electrical properties of Sr and Ca co-doped $BaTiO_3$ nanoparticles. *Ceram. Int.* **2020**, *46 Part A*, 10270–10278. [CrossRef]
21. Morrison, F.D.; Sinclair, D.C.; West, A.R. Doping mechanisms and electrical properties of La-doped $BaTiO_3$ ceramics. *Int. J. Inorg. Mater.* **2001**, *3*, 1205–1210. [CrossRef]
22. Paunović, V.; Prijić, Z.; Dordević, M.; Mitić, V. Enhanced dielectric properties in La modified barium titanate ceramics. *Facta Univ. Ser. Electron. Energet.* **2019**, *32*, 179–193. [CrossRef]
23. Yu, D.; Jin, A.-M.; Zhang, Q.-L.; Yang, H.; Hu, L.; Cheng, D. Scandium and gadolinium co-doped $BaTiO_3$ nanoparticles and ceramics prepared by sol–gel–hydrothermal method: Facile synthesis, structural characterization and enhancement of electrical properties. *Powder Technol.* **2015**, *283*, 433–439. [CrossRef]
24. Oliveira, M.A.; Silva, M.C.O.; M'peko, J.C.; Hernandes, A.C.; MendezGonzález, Y.; Guerra, J.D.S. Investigation of the structural and microstructural properties of Gd^{3+}-modified $BaTiO_3$ ceramics. *Ferroelectrics* **2019**, *545*, 47–54. [CrossRef]
25. Kang, D.W.; Park, T.G.; Kim, J.W.; Kim, J.S.; Lee, H.S.; Cho, H. Effect of Dysprosium Oxide Addition on the Microstructure and Dielectric Properties of $BaTiO_3$ Ceramics. *Electron. Mater. Lett.* **2010**, *6*, 145–149. [CrossRef]
26. Mahapatra, A.K.; Badapanda, T.; Sarangi, S. Investigation of Structural, Dielectric, and Optical Behaviour of Dysprosium-Doped Barium Titanate Ceramics. *ECS J. Solid State Sci. Technol.* **2021**, *10*, 093003. [CrossRef]
27. Tasneem, M.; Kamakshi, K. Dysprosium doping induced effects on structural, dielectric, energy storage density, and electro-caloric response of lead-free ferroelectric barium titanate ceramics. *Electron. Mater.* **2024**, *59*, 1472–1485. [CrossRef]
28. Bhoobash, S.B.; Pradhan, N.; Behera, C. Tuning electrical properties of BaTiO3 with iron modification. *Ceram. Int.* **2023**, *49*, 30076–30089. [CrossRef]
29. Rajan, S.; Mohammed Gazzali, P.M.; Chandrasekaran, G. Impact of Fe on structural modification and room temperature magnetic ordering in $BaTiO_3$. *Spectrochim. Acta Part A Mol. Biomol. Spectrosc.* **2017**, *171*, 80–89. [CrossRef]
30. Gheorghiu, F.; [illegible], M.; Ciomaga, C.E.; Airimioaei, M.; Kalendra, V.; Banys, J.; Dobromir, M.; Tascu, S.; Mitoseriu, L. Preparation and structural characterization of Fe-doped $BaTiO_3$ diluted magnetic ceramics. *Ceram. Int.* **2017**, *43*, 9998–10003. [CrossRef]
31. Yu, Z.; Ang, C.; Guo, R.; Bhalla, A.S. Piezoelectric and strain properties of ceramics. *J. Am. Ceram. Soc.* **2002**, *85*, 1489–1493.
32. Yu, Z.; Ang, C.; Guo, R.; Bhalla, A.S. Dielectric properties of $Ba(Ti_{1-x}Zr_x)O_3$ solid solutions. *Mater. Lett.* **2007**, *61*, 326–329. [CrossRef]
33. Zhu, X.; Zhang, W.; Chen, X.M. Enhanced dielectric and ferroelectric characteristics in Ca-modified $BaTiO_3$ ceramics. *AIP Adv.* **2013**, *3*, 081104. [CrossRef]
34. Jiang, X.; Hao, H.; Yang, Y.; Zhou, E.; Zhang, S.; Wei, P.; Cao, M.; Yao, Z.; Liu, H. Structure and enhanced dielectric temperature stability of $BaTiO_3$-based ceramics by Ca ion B site-doping. *J. Mater.* **2020**, *7*, 259–268. [CrossRef]
35. Li, L.; Fu, R.; Liao, Q.; Ji, L. Doping behaviors of NiO and Nb_2O_5 in $BaTiO_3$ and dielectric properties of $BaTiO_3$-based X7R ceramics. *Ceram. Int.* **2012**, *38*, 1915–1920. [CrossRef]
36. Wang, Y.; Miao, K.; Wang, W.; Qin, Y. Fabrication of lanthanum doped $BaTiO_3$ finegrained ceramics with a high dielectric constant and temperature-stable dielectric properties using hydro-phase method at atmospheric pressure. *J. Eur. Ceram. Soc.* **2017**, *37*, 2385–2390. [CrossRef]
37. Buscaglia, M.T.; Buscaglia, V.; Viviani, M.; Nann, P.; Hanuskova, M. Infuence of foreign ions on the crystal structure of $BaTiO_3$. *J. Eur. Ceram. Soc.* **2000**, *20*, 1997–2007. [CrossRef]

38. Ren, P.; Wang, Q.; Wang, X.; Wang, L.; Wang, J.; Fan, H.; Zhao, G. Effects of doping sites on electrical properties of yttrium doped $BaTiO_3$. *Mater. Lett.* **2016**, *174*, 197–200. [CrossRef]
39. Sakabe, Y.; Wada, N.; Hiramatsu, T.; Tonogaki, T. Dielectric Properties of Fine-Grained $BaTiO_3$ Ceramics Doped with CaO. *Jpn. J. Appl. Phys.* **2002**, *41*, 6922–6925. [CrossRef]
40. Dhahri, A.; Belkahla, A.; Laifi, J.; Gouadria, S.; Elhadi, M.; Dhahri, J.; Dhahri, E. Crystal structure and dielectric properties of the Ca/Y co-substituted $BaTiO_3$. *Inorg. Chem. Commun.* **2022**, *141*, 109570. [CrossRef]
41. Basumatary, R.K.; Singha, K.K.; Sen, S.; Parida, B.N.; Ganesh, M.D.; Pamu, D.; Srivastava, S.K.; Brahma, R. Effect of Co doped BCT on structural, microstructural, dielectric, and multiferroic properties. *Ceram. Int.* **2024**, *50*, 36306–36319. [CrossRef]
42. Chen, X.M.; Chen, X.M.; Wang, T.; Li, J. Dielectric characteristics and their field dependence of (Ba, Ca)TiO_3 ceramics. *Mater. Sci. Eng. B* **2004**, *113*, 117–120. [CrossRef]
43. Fu, D.; Itoh, M.; Koshihara, S. Crystal growth and piezoelectricity of $BaTiO_3$–$CaTiO_3$ solid solution. *Appl. Phys. Lett.* **2008**, *93*, 012904. [CrossRef]
44. Fu, D.; Itoh, M.; Koshihara, S. Invariant lattice strain and polarization in $BaTiO_3$–$CaTiO_3$ ferroelectric alloys. *J. Phys. Condens. Matter* **2010**, *22*, 052204. [CrossRef] [PubMed]
45. Kadira, L.; Elmesbahi, A.; Sayouri, S. Dielectric study of calcium doped barium titanate $Ba_{1-x}Ca_xTiO_3$ ceramics. *Int. J. Phys. Sci.* **2016**, *11*, 71–79.
46. Dash, S.; Mohanty, H.S.; Bhoi, K.; Kant, R.; Kumar, A.; Thomas, R.; Pradhan, D.K. Sintering dependent Ca^{2+} solubility in barium titanate synthesized by sol-gel auto combustion method. *J. Mater. Sci. Mater. Electron.* **2018**, *29*, 20820–20831. [CrossRef]
47. Sampaio, D.V.; Silva, M.S.; Souza, N.R.S.; Santos, J.C.A.; Rezende, M.V.S.; Sil-va, R.S. Electrical characterization of $BaTiO_3$ and $Ba_{0.77}Ca_{0.23}TiO_3$ ceramics synthesized by the proteic sol-gel method. *Ceram. Int.* **2018**, *44*, 15526–15530. [CrossRef]
48. Li, L.Y.; Tang, X.G. Effect of electric field on the dielectric properties and fer-roelectric phase transition of sol-gel derived $(Ba_{0.90}Ca_{0.10})TiO_3$ ceramics. *Mater. Chem. Phys.* **2009**, *115*, 507–511. [CrossRef]
49. Valdez-Nava, Z.; Guillemet-Fritsch, S.; Tenailleau, C.; Lebey, T.; Durand, B.; Chane-Ching, J.Y. Colossal dielectric permittivity of $BaTiO_3$-based nanocrystalline ceramics sintered by spark plasma sintering. *J. Electroceram.* **2009**, *22*, 238–244. [CrossRef]
50. Ctibor, P.; Sedláček, J.; Straka, L.; Lukáč, F.; Neufuss, K. Dielectric Spectroscopy of Calcium Titanate Processed by Spark Plasma Sintering. *Materials* **2023**, *16*, 975. [CrossRef]
51. Choudhury, S.; Bhuiyan, M.A.; Hoque, S.M. Effect of Sintering Temperature on Apparent Density and Transport Properties of $NiFe_2O_4$: Synthesized from Nano Size Powder of NiO and Fe_2O_3. *Int. Nano Lett.* **2011**, *1*, 111–116.
52. Wang, W.; Fu, Z.; Wang, H.; Yuan, R. Influence of Hot Pressing Sintering Temperature and Time on Microstructure and Mechanical Properties of TiB_2 Ceramics. *J. Eur. Ceram. Soc.* **2002**, *22*, 1045–1049. [CrossRef]
53. Mishra, P.; Sonia; Kumar, P. Effect of sintering temperature on dielectric, piezoelectric and ferroelectric properties of BZT–BCT 50/50 ceramics. *J. Alloys Compds.* **2012**, *545*, 210–215. [CrossRef]
54. Chen, T.; Chu, S.; Juang, Y. Effects of sintering temperature on the dielectric and piezoelectric properties of Sr additive Sm-modified $PbTiO_3$ ceramics. *Sens. Actuators A Phys.* **2002**, *102*, 6–10. [CrossRef]
55. Li, C.-X.; Yang, B.; Zhang, S.-T.; Zhang, R.; Cao, W. W. Effects of Sintering Temperature and Poling Conditions on the Electrical Properties of $Ba_{0.70}Ca_{0.30}TiO_3$ Diphasic Piezoelectric Ceramics. *Ceram. Int.* **2013**, *39*, 2967–2973. [CrossRef]
56. Tang, X.G.; Wang, J.; Wang, X.X.; Chan, H.L.W. Effects of grain size on the dielectric properties and tunabilities of sol-gel derived $Ba(Zr_{0.2}Ti_{0.8})O_3$ ceramics. *Solid State Commun.* **2004**, *131*, 163–168. [CrossRef]
57. Kingery, W.D.; Bowen, H.K.; Uhlmann, D.R. *Introduction to Ceramics*; Wiley: New York, NY, USA, 1976; p. 1056.
58. Tullis, J.; Yund, R.A. Grain Growth Kinetics of Quartz and Calcite Aggregates. *J. Geol.* **1982**, *90*, 301–318. [CrossRef]
59. Barsoum, M.W. *Fundamentals of Ceramics*; McGraw Hill: New York, NY, USA, 1997; p. 668.
60. Arlt, G.; Hennings, D.; de With, G. Dielectric properties of fine-grained barium titanate ceramics. *J. Appl. Phys.* **1985**, *58*, 1619–1625. [CrossRef]
61. Sato, Y.; Kanai, H.; Yamashita, Y. Large electromechanical coupling factors in perovskite binary material system. *Jpn. J. Appl. Phys.* **1994**, *33*, 1383.
62. Yamamoto, T. Optimum preparation methods for piezoelectric ceramics and their evaluation. *Am. Ceram. Soc. Bull.* **1992**, *71*, 978–985.
63. Singh, A.K.; Goel, T.C.; Mendiratta, R.G.; Thakur, O.P.; Prakash, C. Dielectric Properties of Mn-Substituted Ni–Zn Ferrites. *J. Appl. Phys.* **2002**, *91*, 6626–6629. [CrossRef]
64. Arlt, G. The influence of microstructure on the properties of ferroelectric ceramics. *Ferroelectrics* **1990**, *104*, 217. [CrossRef]
65. Curecheriu, L.; Buscaglia, M.T.; Buscaglia, V.; Zhao, Z.; Mitoseriu, L. Grain size effect on the nonlinear dielectric properties of barium titanate ceramics. *Appl. Phys. Lett.* **2010**, *97*, 242909–2429093. [CrossRef]
66. Zhao, Z.; Buscaglia, V.; Viviani, M.; Buscaglia, M.T.; Mitoseriu, L.; Testino, A.; Nygren, M.; Johnsson, M.; Nanni, P. Grain-size effects on the ferroelectric behavior of dense nanocrystalline $BaTiO_3$ ceramics. *Phys. Rev. B* **2004**, *70*, 024107–0241078. [CrossRef]
67. Buessem, W.R.; Kahn, M. Effects of Grain Growth on the Distribution of Nb in $BaTiO_3$ Ceramics. *J. Am. Ceram. Soc.* **1971**, *54*, 491–495. [CrossRef]
68. Ku Muhsen, K.N.D.; Osman, R.A.M.; Idris, M.S.; Nadzri, N.I.M.; Jumali, M.H.H. Effect of sintering temperature on $(Ba_{0.85}Ca_{0.15})(Sn_xZr_{0.1-x}Ti_{0.9})O_3$ for piezoelectric energy harvesting applications. *Ceram. Int.* **2021**, *47*, 13107–13117. [CrossRef]

69. Stanculescu, R.; Ciomaga, C.E.; Padurariu, L.; Galizia, P.; Horchidan, N.; Capiani, C.; Galassi, C.; Mitoseriu, L. Study of the role of porosity on the functional properties of (Ba,Sr)TiO_3 ceramics. *J. Alloys Compd.* **2015**, *643*, 79–87. [CrossRef]
70. Faggio, G.; Politano, G.G.; Lisi, N.; Capasso, A.; Messina, G. The structure of chemical vapor deposited graphene substrates for graphene-enhanced Raman spectroscopy. *J. Phys. Condens. Matter.* **2004**, *36*, 195303. [CrossRef]
71. Das, P.; Bathula, S.; Gollapudi, S. Evaluating the effect of grain size distribution on thermal conductivity of thermoelectric materials. *Nano Express* **2020**, *1*, 020036. [CrossRef]
72. Zhang, X.; Chen, G.; Fu, C.; Wang, F. Effects of sintering temperature and holding time on the microstructure and electric properties of $Ba(Zr_{0.3}Ti_{0.7})O_3$ ceramics. *Process. Appl. Ceram.* **2018**, *12*, 45–55. [CrossRef]
73. Ganguly, P.; Jha, A.K. Structural and Electrical Properties of $Ba_{5-x}Ca_xSmTi_3Nb_7O_{30}$ (x = 0–5) Ferroelectric Ceramics. *J. Alloys Compd.* **2010**, *495*, 7–12. [CrossRef]
74. Rajan, S.; Mohammed Gazzali, P.M.; Chandrasekaran, G. Electrical and Magnetic Phase Transition Studies of Fe and Mn Co-Doped $BaTiO_3$. *J. Alloys Compd.* **2016**, *656*, 98–109. [CrossRef]
75. Vittayakorn, N.; Charoonsuk, T.; Kasiansin, P.; Boonchom, B. Dielectric Properties and Phase Transition Behaviors in $(1-x)PbZrO_{3-x}Pb(Mg_{1/2}W_{1/2})O_3$ Ceramics. *J. Appl. Phys.* **2009**, *106*, 064104. [CrossRef]
76. Kartashev, A.V.; Bondarev, V.S.; Flerov, I.N.; Gorev, M.V.; Pogorel'tsev, E.I.; Shabanov, A.V.; Molokeev, M.S.; Guillemet-Fritsch, S.; Raevskii, I.P. Study of the Physical Properties and Electrocaloric Effect in the $BaTiO_3$ Nano- and Microceramics. *Phys. Solid State* **2019**, *61*, 1052–1061. [CrossRef]
77. Levanyuk, A.P.; Osipov, V.V.; Sigov, A.S.; Sobyanin, A.A. Change of defect structure and the resultant anomalies in the properties of substances near phase-transition points. *J. Exp. Theor. Phys.* **1979**, *49*, 176.
78. Dubrovskii, I.M.; Krivoglaz, M.A. Phase Transitions of the Second Kind in Crystals Containing Dislocations. *Sov. Phys. JETP* **1979**, *50*, 512–520.
79. Kallaev, S.N.; Omarov, Z.M.; Bakmaev, A.G.; Abdulvakhidov, K. Heat Capacity of Nanostructured $BaTiO_3$ Ceramics. *Phys. Solid State* **2013**, *55*, 1095–1097. [CrossRef]
80. Yoon, S.-H.; Randall, C.A.; Hur, K.-H. Correlation Between Resistance Degradation and Thermally Stimulated Depolarization Current in Acceptor (Mg)-Doped $BaTiO_3$ Sub-Micrometer Fine-Grain Ceramics. *J. Am. Ceram. Soc.* **2010**, *93*, 593–599. [CrossRef]

materials

MDPI

Article

Preparation and Properties of Nb^{5+}-Doped BCZT-Based Ceramic Thick Films by Scraping Process

Yang Zou [1], Bijun Fang [1,*], Xiaolong Lu [1], Shuai Zhang [1] and Jianning Ding [1,2,*]

1 School of Materials Science and Engineering, Jiangsu Collaborative Innovation Center of Photovoltaic Science and Engineering, Jiangsu Province Cultivation Base for State Key Laboratory of Photovoltaic Science and Technology, National Experimental Demonstration Center for Materials Science and Engineering, Changzhou University, Changzhou 213164, China; s22010856091@smail.cczu.edu.cn (Y.Z.); xllu@cczu.edu.cn (X.L.); shuaizhang@cczu.edu.cn (S.Z.)

2 School of Mechanical Engineering, Yangzhou University, Yangzhou 225127, China

* Correspondence: fangbj@cczu.edu.cn (B.F.); dingjn@cczu.edu.cn (J.D.); Tel./Fax: +86-519-86330095 (B.F.)

Abstract: A bottleneck characterized by high strain and low hysteresis has constantly existed in the design process of piezoelectric actuators. In order to solve the problem that actuator materials cannot simultaneously exhibit large strain and low hysteresis under relatively high electric fields, Nb^{5+}-doped $0.975(Ba_{0.85}Ca_{0.15})[(Zr_{0.1}Ti_{0.9})_{0.999}Nb_{0.001}]O_3$-$0.025(Bi_{0.5}Na_{0.5})ZrO_3$ ($BCZTNb_{0.001}$-0.025BiNZ) ceramic thick films were prepared by a film scraping process combined with a solid-state twin crystal method, and the influence of sintering temperature was studied systematically. All $BCZTNb_{0.001}$-0.025BiNZ ceramic thick films sintered at different sintering temperatures have a pure perovskite structure with multiphase coexistence, dense microstructure and typical dielectric relaxation behavior. The conduction mechanism of all samples at high temperatures is dominated by oxygen vacancies confirmed by linear fitting using the Arrhenius law. As the sintering temperature elevates, the grain size increases, inducing the improvement of dielectric, ferroelectric and field-induced strain performance. The 1325 °C sintered $BCZTNb_{0.001}$-0.025BiNZ ceramic thick film has the lowest hysteresis (1.34%) and relatively large unipolar strain (0.104%) at 60 kV/cm, showing relatively large strain and nearly zero strain hysteresis compared with most previously reported lead-free piezoelectric ceramics and presenting favorable application prospects in the actuator field.

Keywords: BCZT based ceramic thick films; film scraping process; sintering condition; conduction mechanism; low hysteresis strain

1. Introduction

At present, lead-based piezoelectric ceramics with excellent field-induced strain performance occupy a dominant position in the application field of actuators [1–3]. However, many countries around the world have advocated for environmental protection in recent years [4]. Due to the toxicity of PbO in raw materials, the preparation and application of lead-based ceramics are constrained, thus it is urgently necessary to develop lead-free ceramic materials with excellent electrostrain performance to gradually replace lead-based ceramics [4–6].

In 2009, Ren et al. initially proposed $(Ba_{0.85}Ca_{0.15})(Zr_{0.1}Ti_{0.9})O_3$ (BCZT) lead-free piezoelectric ceramics with $BaTiO_3$ as the matrix [7]. The unipolar strain ($S_{unipolar}$) and the converse piezoelectric coefficient (d_{33}^*) values of BCZT ceramics were astonishingly 0.057% and 1140 pm/V under an electric field of 5 kV/cm, respectively, which attracted extensive attention from scholars both domestically and internationally. Unfortunately, the BCZT ceramics exhibited a high strain hysteresis (Hys) exceeding 20% at 5 kV/cm, which significantly restricted their applications in the actuator field. This limitation arises from the requirement for actuator materials to simultaneously possess high strain and low strain hysteresis [8,9]. Consequently, the significant challenge lies in enabling materials to maintain relatively large strain while simultaneously exhibiting extremely low strain hysteresis [9–11].

According to reports in the literature, the introduction of a second component to form a pseudo-binary solid solution in the piezo-materials can cause disorder in the composition and enhance the random field, disrupting the long-range ordered ferroelectric states within the materials and leading to the effective generation of polar nanoregions (PNRs) [12,13], which suppresses the strain hysteresis in materials due to their strong response to the external electric field [14,15]. For instance, Tian et al. incorporated the second component $BaZrO_3$ into $Na_{0.52}K_{0.48}Nb_{0.9}Sb_{0.1}O_3$ to form a pseudo-binary solid solution of $(100 - x)Na_{0.52}K_{0.48}Nb_{0.9}Sb_{0.1}O_3$-$xBaZrO_3$ [16]. As x increased from 0 to 4, the presence of PNRs caused the Hys value at 100 kV/cm to decrease from approximately 40% to 7.4%, while also exhibiting a relatively high strain of 0.2% [16]. Huang et al. introduced $KNbO_3$ into $BaTiO_3$ to create solid solution of $(1 - x)BaTiO_3$-$xKNbO_3$ [17]. Similarly driven by the presence of PNRs, the $0.94BaTiO_3$-$0.06KNbO_3$ ceramics achieved an extremely low Hys of approximately 3% and a relatively high $S_{unipolar}$ of approximately 0.10% at 80 kV/cm [17].

Additionally, it is widely recognized that soft donor doping can promote the rotation and switching of ferroelectric domains by creating cationic vacancies through high-valence ion substitution for low-valence ion, thereby reducing lattice stress and enhancing field-induced strain [18–20]. Simultaneously, the soft donor ion doping also leads to structural disorder and an enhanced random field, disrupting long-range ordered ferroelectric states and inducing the formation of PNRs within the piezo-materials, consequently reducing the strain hysteresis [19,21,22]. For example, Habib et al. reported the modification of $0.70Bi_{1.03}FeO_3$-$0.30BaTiO_3$ through A-site La^{3+} donor doping to obtain $0.70Bi_{1.03}FeO_3$-$0.30Ba_{(1-x)}La_xTiO_3$ ceramics [22]. As the x increased from 0 to 0.015, the $S_{unipolar}$ value at 55 kV/cm rose from 0.150% to 0.188%, while the Hys value decreased from over 30% to approximately 20% [22]. Kang et al. modified $0.67BiFeO_3$-$0.33BaTiO_3$ with B-site Sb^{5+} donor doping to obtain $0.67BiFe_{1-x}Sb_xO_3$-$0.33BaTiO_3$ ceramics [19]. At 50 kV/cm, when x was 0.25%, the $S_{unipolar}$ and Hys values of $0.67BiFe_{99.75\%}Sb_{0.25\%}O_3$-$0.33BaTiO_3$ ceramics were 0.168% and 49%, respectively, exhibiting a significant improvement over the $S_{unipolar}$ of 0.120% and the Hys of 59% in the $0.67BiFeO_3$-$0.33BaTiO_3$ ceramics when x was 0 [19].

Therefore, in our previous research, $(Bi_{0.5}Na_{0.5})ZrO_3$ (BiNZ) was used as the secondary component to introduce into BCZT to form the solid solution $(1 - x)(Ba_{0.85}Ca_{0.15})(Zr_{0.1}Ti_{0.9})O_3$-$x(Bi_{0.5}Na_{0.5})ZrO_3$ (BCZT-xBiNZ), where the orbital hybridization between the 6s orbital of Bi^{3+} and the 2p orbital of O^{2-} will result in high maximum polarization [23,24]. As we all know, the sintering temperature of preparing pure BCZT is comparably high (~1450 °C or above) [7], causing significant energy consumption [5]. Nevertheless, the low melting point characteristic of BiNZ can effectively reduce the sintering temperature and the energy consumption during the sintering process [23]. Via designing composition and tailoring sintering temperature, 1325 °C sintered BCZT-0.025BiNZ ceramics exhibited a higher relative density of 93.09% and optimal field-induced strain performance of $S_{unipolar}$ 0.055% with low Hys of 6.55% at 20 kV/cm. In order to further improve the field-induced strain performance, B-site Nb^{5+} donor doping was undertaken to form $0.975(Ba_{0.85}Ca_{0.15})[(Zr_{0.1}Ti_{0.9})_{1-x}Nb_x]O_3$-$0.025(Bi_{0.5}Na_{0.5})ZrO_3$ ($BCZTNb_x$-0.025BiNZ) ceramics. When x = 0.001, the $S_{unipolar}$ value of $BCZTNb_{0.001}$-0.025BiNZ ceramics sintered at 1325 °C reached 0.058% with a significantly reduced Hys of 3.65% at 20 kV/cm, presenting a marked improvement in field induced strain performance.

As is widely recognized, traditional block ceramics are relatively thick, resulting in a greater number of defects and decreased dielectric breakdown strength under high external electric field compared with thinner ceramics [25]. While grinding and polishing manually or by machine can reduce the thickness of ceramics, extended preparation time is required and increased mechanical breakage is possible, which is not conducive to industrial production [25,26]. In 2020, Yan et al. successfully fabricated $0.75Bi_{0.58}Na_{0.42}TiO_3$-$0.25SrTiO_3$ ceramic films with a thickness of ~0.04 mm using the film scraping process, which increased the maximum breakdown strength to 569 kV/cm [27]. Such results demonstrate that the film scraping process not only simplifies the fabrication process of thinner ceramics but also significantly enhances the maximum withstanding external electric field capability [25–27].

Furthermore, the preparation and processing of ceramic materials are closely intertwined with their structure and performance [23,28], among which the selection of sintering temperature is of paramount importance [23]. The sintering temperature directly influences the grain size and its uniformity in ceramics, consequently impacting their relative density and electrical properties [29,30]. It is worth noting that grain size is a significant factor influencing the strain performance of ceramics, as it affects the motion and switching of ferroelectric domains [19,31,32]. In the BCZT-based ceramics, the strain primarily originates from the switching of ferroelectric domains [33]. For example, Cai et al. investigated the sintering of BCZT ceramics in the temperature range of 1350 °C to 1500 °C and observed an increase in the grain size from 8.28 μm (1350 °C) to 44.37 μm (1500 °C) [33]. Correspondingly, the $S_{unipolar}$ value of the BCZT ceramics increased from 0.086% (1350 °C) to 0.115% (1500 °C) at 20 kV/cm [33]. However, excessively high sintering temperatures are undesirable as they can lead to abnormal grain growth, increased grain size non-uniformity, greater surface pores in the microstructure, resulting in decreased relative density and consequently diminished material performance [31]. For example, Bijalwan et al. introduced 0.07 wt% CeO_2 into BCZT, in which when the sintering temperature increased from 1275 °C to 1350 °C, the grain size increased from ~5 μm to ~13 μm, resulting in the increase in bulk density from ~5 g/cm^3 to 5.64 g/cm^3, maximum dielectric constant (ε_m) at 1 kHz from ~2500 to ~4091 and piezoelectric constant (d_{33}) from 50 pC/N to 507 pC/N after poling at 30 kV/cm; whereas further raising the sintering temperature to 1450 °C, the grain size of ceramics increased to over 25 μm, while bulk density, ε_m and d_{33} decreased to below 5.6 g/cm^3, ~3500 and ~200 pC/N, respectively [31]. Hence, it is imperative to investigate the influence of sintering temperature on the structure and properties of ceramic materials.

Thus, this study employed the film scraping process to prepare the BCZTNb$_{0.001}$-0.025BiNZ ceramic thick films, and the influence of sintering temperature on their structure and properties under relatively high external electric field was investigated. At 60 kV/cm, the relatively high $S_{unipolar}$ (0.104%) and near-zero Hys (1.34%) were achieved in the BCZTNb$_{0.001}$-0.025BiNZ ceramic thick film sintered at 1325 °C, exhibiting promising application prospects in actuator technology.

2. Experimental Procedure

2.1. Sample Preparation

The schematic of preparing ceramic thick films by scraping process is shown in Figure 1. The $0.975(Ba_{0.85}Ca_{0.15})[(Zr_{0.1}Ti_{0.9})_{0.999}Nb_{0.001}]O_3$-$0.025(Bi_{0.5}Na_{0.5})ZrO_3$ (BCZTNb$_{0.001}$-0.025BiNZ) ceramic thick films were prepared by a film scraping process [25,26], in which the BCZTNb$_{0.001}$-0.025BiNZ powder was prepared by the solid-state twin crystal method, that was, the BCZTNb$_{0.001}$ powder and BiNZ powder were synthesized separately [29]. All raw materials were commercially sold high purity oxides and carbonates from Sinopharm Group Chemical Reagent Co., Ltd. (Shanghai, China) without further processing and dried completely at 120 °C before preparation. Stoichiometrically weighed $BaCO_3$, $CaCO_3$, ZrO_2, TiO_2 and Nb_2O_5 were mixed and ground uniformly and calcined at 1225 °C for 4 h to prepare the BCZTNb$_{0.001}$ powder. Similarly, thorough grinding of stoichiometrically weighed Bi_2O_3, Na_2CO_3 and ZrO_2 was calcined at 625 °C for 3 h to obtain the BiNZ powder.

To prepare the BCZTNb$_{0.001}$-0.025BiNZ ceramic thick films, accurately weighed stoichiometric BCZTNb$_{0.001}$ powder and BiNZ powder were mixed and ground homogeneously in anhydrous ethanol, and then were sieved through a 150-mesh sieve. An appropriate quantity of polyvinyl butyral was added to the sieved mixture, and they were further ground until achieving uniformity. Subsequently, an appropriate quantity of n-methylpyrrolidone and dibutyl phthalate was added, followed by further grinding until the casting slurry was formed. The slurry was scraped onto a polyester release film with a scraper to form the thick films with a thickness of 0.75 mm. The scraped thick films were placed in an 80 °C oven for 15 h and then cut into circular film plates with a diameter of 12 mm. The organic compounds in film plates were separately removed at 600 °C for 10 h. The decarbonated films were sintered at 1305 °C–1345 °C for two hours to obtain ceramic thick films. Prior to sintering, the decarbonated films were covered with powder of the same composition in a covered alumina crucible to reduce the volatilization of Bi and Na. After sintering at different sintering temperatures, the $BCZTNb_{0.001}$-0.025BiNZ ceramic thick films with a thickness of approximately 0.25 mm were obtained.

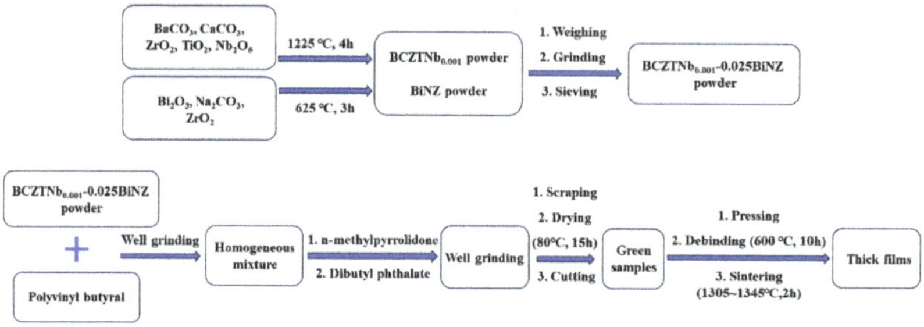

Figure 1. Schematic of preparing ceramic thick films by scraping process.

2.2. Materials Characterization

The crystal structure, phase structure, microstructure and grain size distribution of $BCZTNb_{0.001}$-0.025BiNZ ceramic thick films were analyzed using X-ray diffraction method (Rigaku D/max-2500/PC, $CuK\alpha_1$, λ = 1.5406 Å, Japan), GSAS 1 software, JSM-IT100 scanning electron microscopy (SEM) and nano-measurement software, respectively. In order to further evaluate their electrical properties, the ceramic thick films were coated with silver paste on both sides and fired at 650 °C for 0.5 h. The dielectric properties from room temperature to 180 °C at 100 Hz–2 MHz and complex impedance spectra at high temperatures (390 °C–550 °C or 410 °C–550 °C) at 100 Hz–1 MHz were measured by Partulab HDMS-1000 system. The ferroelectric properties (P-E) at 60 kV/cm and 1 Hz, as well as the field-induced strain (S-E) properties at 60 kV/cm and 10 Hz, were tested by a ferroelectric analyzer (Precision LC II, Radiant Technologies Inc., Albuquerque, NM, USA). Before testing the ferroelectric and field-induced strain properties, the ceramic thick films were immersed in silicone oil at a temperature of 80 °C to 100 °C for 5 min to mitigate the risk of breakdown during the testing process.

3. Results and Discussion

3.1. Phase Structure and Refinement

Figure 2a shows XRD patterns of the $BCZTNb_{0.001}$-0.025BiNZ ceramic thick films sintered at 1305 °C to 1345 °C. All $BCZTNb_{0.001}$-0.025BiNZ ceramic thick films exhibit a pure perovskite phase, indicating the complete integration of Nb^{5+} ion into the crystal lattice of BCZT-0.025BiNZ and forming a stable solid solution. Figure 2b shows locally amplified XRD patterns of the $BCZTNb_{0.001}$-0.025BiNZ films, clearly demonstrating that there is no splitting in the (200) diffraction peak. Whereas the {100} diffraction reflection presents splitting into both the (100) and (001) peaks as evidently shown in Figure 2a, and the (200) peak displays broadening although without obvious splitting. Such phenomena suggest multiple phases coexist within the $BCZTNb_{0.001}$-0.025BiNZ ceramic thick films [29], and the structure can generally be regarded as a pseudo-cubic structure [34,35].

Detailed multiple phases coexistence was analyzed by XRD Rietveld refinement as shown in Figure 3 of the $BCZTNb_{0.001}$-0.025BiNZ ceramic thick films sintered at different sintering temperatures. The fitted R_{wp} and χ^2 values for all ceramic thick films are below 8.9% and 3.6%, respectively, indicating a high degree of accuracy in the fitting results, and then, the XRD Rietveld refinement parameters are presented in Table 1. From analyzing Figure 3 and Table 1, it is evident that all ceramic thick films exhibit the presence of tetragonal, rhombohedral and orthorhombic phases simultaneously, presenting a similar phase structure to the pure BCZT ceramics reported by Ren et al. [7]. Notably, as the sintering temperature increases from 1305 °C to 1345 °C, the content of rhombohedral phase R3m increases from 8.17% (1305 °C) to 30.51% (1345 °C), demonstrating a pronounced upward trend. The tetragonal phase P4mm content in the 1305 °C and 1325 °C sintered films is comparable (~27%), which decreases significantly to 5.49% with further elevation of the sintering temperature to 1345 °C. With the increase in sintering temperature, the content of the orthorhombic phase Amm2 decreases apparently to 46.18% at the sintering temperature of 1325 °C, and then increases recoverably to a comparable value (~64%). In all ceramic thick films, the phase content of the orthorhombic phase obtained by Rietveld refinement is the highest, which is in line with the coexistence of multiple phases obtained from XRD, and the films can be viewed as a pseudo-cubic structure [34,35].

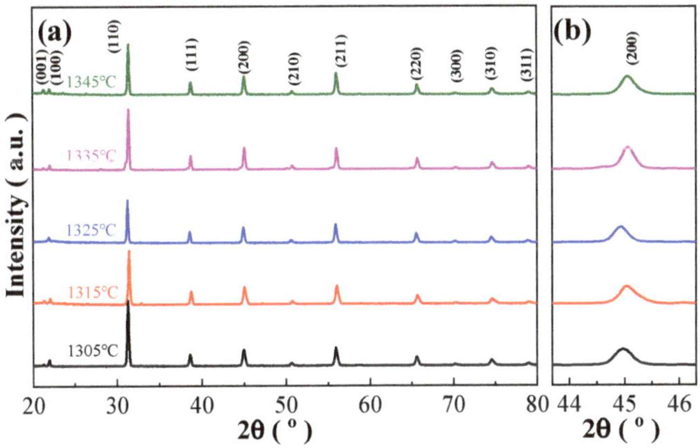

Figure 2. (**a**) XRD patterns of the $BCZTNb_{0.001}$-0.025BiNZ ceramic thick films sintered at different sintering temperatures; (**b**) locally amplified XRD patterns of the $BCZTNb_{0.001}$-0.025BiNZ ceramic thick films around (200) peak.

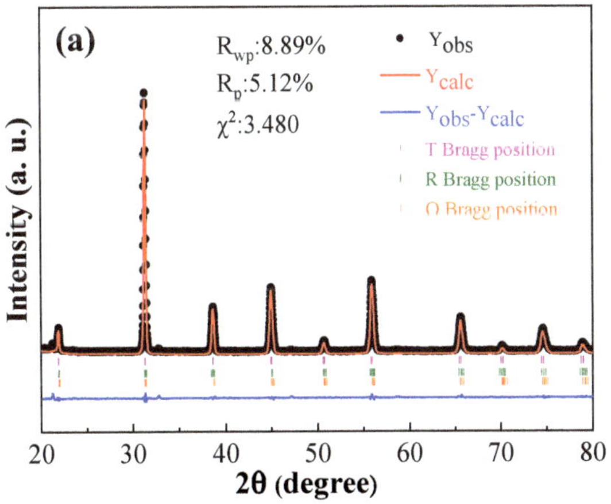

Figure 3. *Cont.*

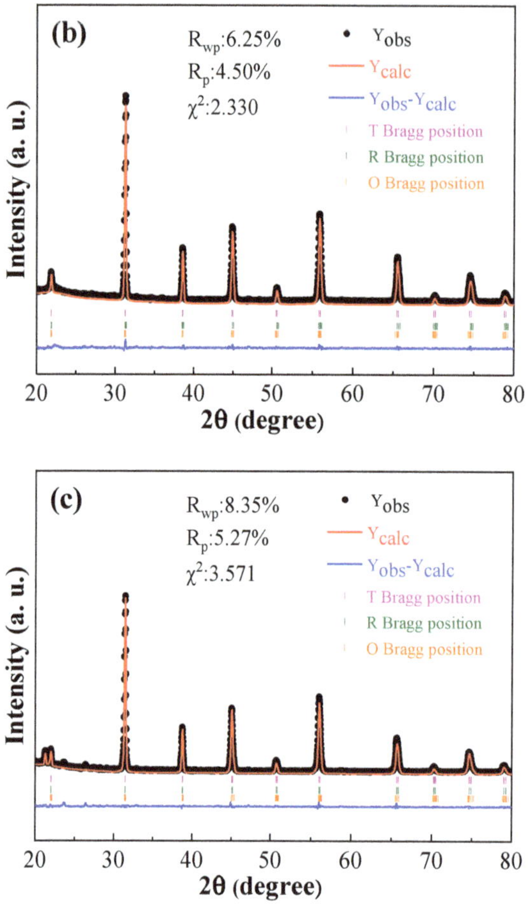

Figure 3. XRD Rietveld refinement results of the $BCZTNb_{0.001}$-0.025BiNZ ceramic thick films sintered at different sintering temperatures: (**a**) 1305 °C; (**b**) 1325 °C; (**c**) 1345 °C.

Table 1. Space group, phase fraction, lattice constants of the $BCZTNb_{0.001}$-0.025BiNZ ceramic thick films sintered at different sintering temperatures.

Sintering Temperature	Space Group	a (Å)	b (Å)	c (Å)	α = β = γ (°)	Fraction (%)
1305 °C	P4mm	4.019066	4.019066	4.015660	90	26.79
	R3m	4.010242	4.010242	4.010242	89.746	8.17
	Amm2	3.994396	5.674225	5.675882	90	65.04
1325 °C	P4mm	4.014551	4.014551	4.013084	90	27.81
	R3m	4.006793	4.006793	4.006793	89.845	26.01
	Amm2	4.001672	5.677942	5.689648	90	46.18
1345 °C	P4mm	4.012797	4.012797	4.008653	90	5.49
	R3m	4.011222	4.011222	4.011222	90.019	30.51
	Amm2	3.992369	5.675968	5.687695	90	64.00

3.2. SEM Images and Grain Size Distribution

Figure 4 shows the SEM images and grain size distribution statistics of the $BCZTNb_{0.001}$-0.025BiNZ ceramic thick films sintered at different sintering temperatures. The grain boundaries of all ceramic thick films are clear, in which the microstructure is dense and has relatively few surface pores. Additionally, the majority of grains are irregular polygons with a minority being nearly circular shape grains, indicating that the solid-state sintering is the primary mechanism during the sintering process, while liquid-phase sintering plays a secondary role [23,29,36]. As the sintering temperature rises, the grain size exhibits a nearly linear increase, and the average grain size of films reach 1.57 μm (1305 °C), 2.18 μm (1325 °C) and 3.13 μm (1345 °C), respectively. In comparison with the grain size of 8.28 μm for the pure BCZT ceramics sintered at 1350 °C reported by Cai et al. [33], the grain size of all ceramic thick film is remarkably decreased, suggesting that the introduction of a second component BiNZ restrains grain growth. When sintered at the lower temperature of 1305 °C, the ceramic thick film displays a smaller grain size with excellent grain distribution, whereas the grain growth is not sufficient, resulting in few surface pores and a lower density. Subsequently, at a higher sintering temperature of 1325 °C, grain growth becomes sufficient while still maintaining good grain distribution uniformity, leading to a denser structure without noticeable surface pores. However, as the sintering temperature is further increased to 1345 °C, the excessive high sintering temperature causes some grains to grow abnormally, resulting in a decrease in the uniformity of grain size. Overall, at an appropriate sintering temperature, the $BCZTNb_{0.001}$-0.025BiNZ ceramic thick films can obtain dense microstructure and high density [25,26]. Additionally, compared with the higher sintering temperature (~1450 °C or above) for preparing pure BCZT ceramics [7], the $BCZTNb_{0.001}$-0.025BiNZ ceramic thick films achieve low-temperature densified sintering at ~1325 °C, greatly reducing energy consumption.

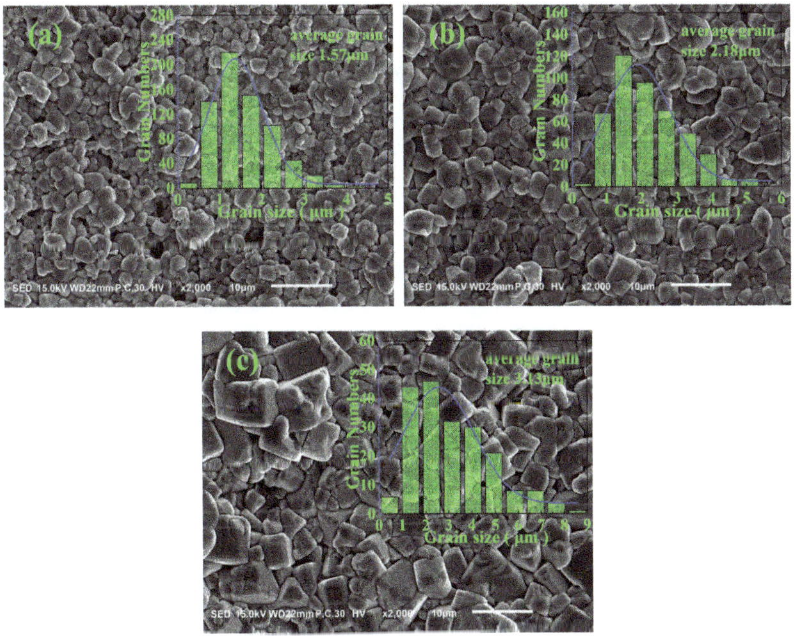

Figure 4. SEM images and grain size statistics (insets) of the $BCZTNb_{0.001}$-0.025BiNZ ceramic thick films sintered at different sintering temperatures: (**a**) 1305 °C; (**b**) 1325 °C; (**c**) 1345 °C.

3.3. Dielectric Properties and Dielectric Response Fitting

Figure 5a,b shows dielectric properties versus temperature relationship of the $BCZTNb_{0.001}$-0.025BiNZ ceramic thick films sintered at different sintering temperatures, as well as curves of the temperature of maximum dielectric constant (T_m) and the maximum dielectric constant (ε_m) with sintering temperature at 10 kHz, respectively. All ceramic thick films exhibit a broad dielectric peak, and dielectric constant tends to increase accompanied by the variation of T_m with elevating sintering temperature, indicating relaxor ferroelectrics characteristic. By further analyzing the two charts, it is observed that T_m initially increases and then decreases with the rising sintering temperature, in which the maximum T_m value of 73 °C is achieved in the 1335 °C sintered ceramic film. Additionally, the ε_m value fluctuates around ~6200 between sintering temperature of 1305 °C–1325 °C, which increases rapidly with further increasing sintering temperature, and reaches the maximum value of 8059.5 in the 1345 °C sintered ceramic film. For comparison, the ε_m and T_m values of the 1450 °C sintered pure BCZT ceramics are ~18,000 and ~93 °C, respectively, reported by Ren et al. [7]. Furthermore, the rapid increase in dielectric properties with elevating sintering temperature can be mainly attributed to the relationship between the dielectric properties of ceramic materials and grain size, in which larger grain size facilitates domain wall movement, resulting in a higher dielectric constant [31,32,37]. This finding aligns with the conclusion drawn from Figure 4. Furthermore, all $BCZTNb_{0.001}$-0.025BiNZ ceramic thick films exhibit a rather low dielectric loss (<0.03), which is especially suitable for electronic industry application.

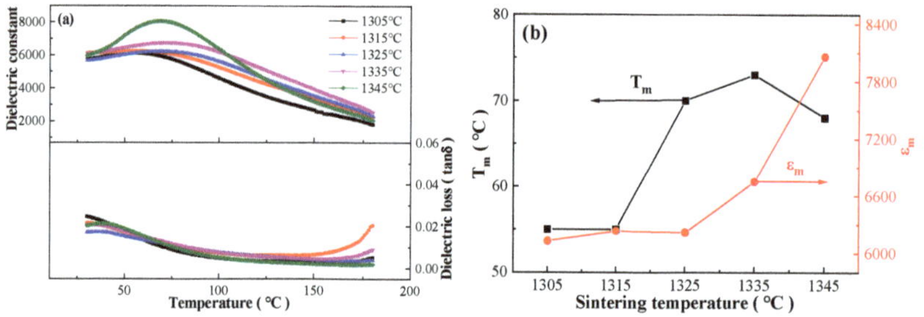

Figure 5. (**a**) Dielectric performance-temperature curves of the $BCZTNb_{0.001}$-0.025BiNZ ceramic thick films sintered at different sintering temperatures at 10 kHz; (**b**) curves of T_m and ε_m with sintering temperature at 10 kHz.

Figure 6 shows the effect of frequency on dielectric properties of the $BCZTNb_{0.001}$-0.025BiNZ ceramic thick films sintered at different sintering temperatures to disclose the relaxor ferroelectrics nature. All ceramic thick films display pronounced frequency dispersion, especially below and around the T_m temperatures, and all dielectric peaks exhibit a broadening characteristic as compared with the sharpening dielectric peak of the normal ferroelectrics such as pure $BaTiO_3$ [17], indicating the typical relaxor ferroelectrics behavior [10,38]. With the increase in sintering temperature, width of dielectric peak, dielectric constant and loss tangent, T_m and ε_m values change accordingly.

At high testing temperatures, all ceramic thick films exhibit a significant increase in dielectric loss particularly at low frequency 100 Hz, possibly attributed to the oxygen vacancy conduction resulting from the evaporation of some metals (Bi and Na) and the heterovalent substitution of Bi^{3+} at the A-site and Nb^{5+} at the B-site within the perovskite structure [39]. The defect equations that generate oxygen vacancies are as follows [40,41]:

$$4Na_{Na}^{\times} + 2O_O^{\times} \rightarrow 4Na\uparrow + O_2\uparrow + 4V_{Na}^{'} + 2V_O^{\cdot\cdot} \quad (1)$$

$$4Bi_{Bi}^{\times} + 6O_O^{\times} \rightarrow 4Bi\uparrow + 3O_2\uparrow + 4V_{Bi}^{'''} + 6V_O^{\cdot\cdot} \quad (2)$$

$$Bi_2O_3 \overset{3(BaCa)O\cdot(ZrTi)O_2}{\rightarrow} 2Bi_{Ba/Ca}^{\cdot} + V_{Ba/ca}^{''} + 3O_O^{\times} + 3(ZrTi)O_2 \quad (3)$$

$$2Nb_2O_5 \overset{5(BaCa)O\cdot(ZrTi)O_2}{\rightarrow} 4Nb_{Zr/Ti}^{\cdot} + V_{Zr/Ti}^{''''} + 10O_O^{\times} + 5(BaCa)O \quad (4)$$

This conduction mechanism belongs to thermally activated conduction with a thermal activation discontinuity, so there will be larger dielectric loss at high temperatures [42], which will be further analyzed in the following complex impedance spectra study.

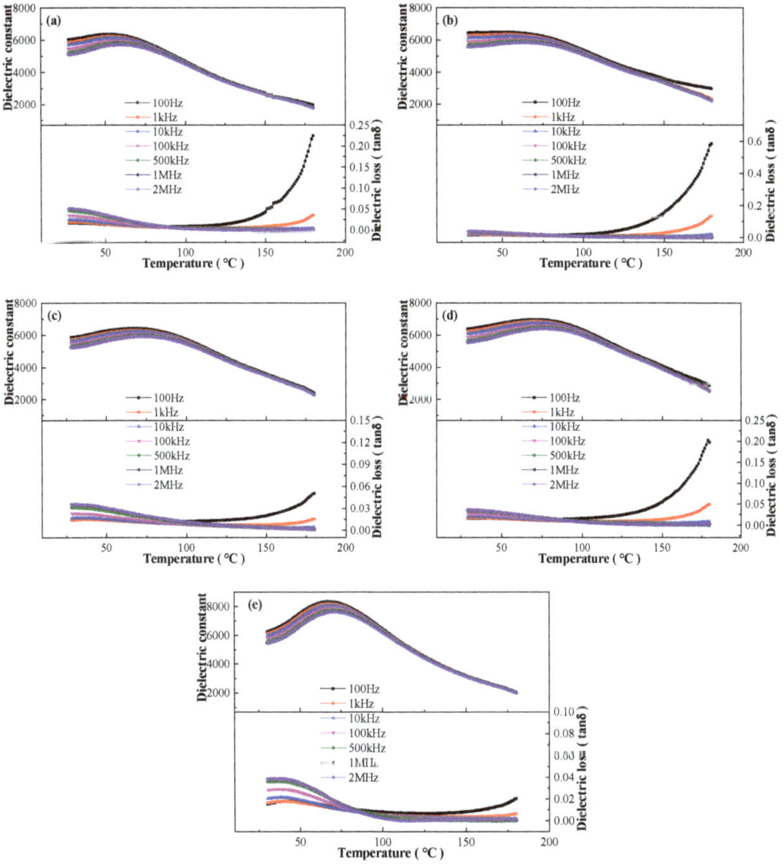

Figure 6. Dielectric performance-temperature curves at several frequencies of the BCZTNb$_{0.001}$-0.025BiNZ ceramic thick films sintered at different sintering temperatures: (**a**) 1305 °C; (**b**) 1315 °C; (**c**) 1325 °C; (**d**) 1335 °C; (**e**) 1345 °C.

The relaxation characteristic of the BCZTNb$_{0.001}$-0.025BiNZ ceramic thick films sintered at various temperatures is further analyzed by the exponential law $\frac{1}{\varepsilon} - \frac{1}{\varepsilon_m} = \frac{(T-T_m)^\gamma}{C'}$ fitting using the dielectric data at 10 kHz [38,41], where C′ is a constant and γ is the dispersion factor, and the results of exponential law fitting are shown in Figure 7. It is evident that all ceramic thick films exhibit nearly typical relaxor ferroelectrics nature and have a significantly large value of the dispersion factor γ of approximately 2, with specific values of 1.980 (1305 °C), 2.146 (1315 °C), 2.008 (1325 °C), 1.977 (1335 °C) and 2.032 (1345 °C), respectively. The above research indicates that the BCZTNb$_{0.001}$-0.025BiNZ ceramic thick films possess a notably high level of relaxation degree.

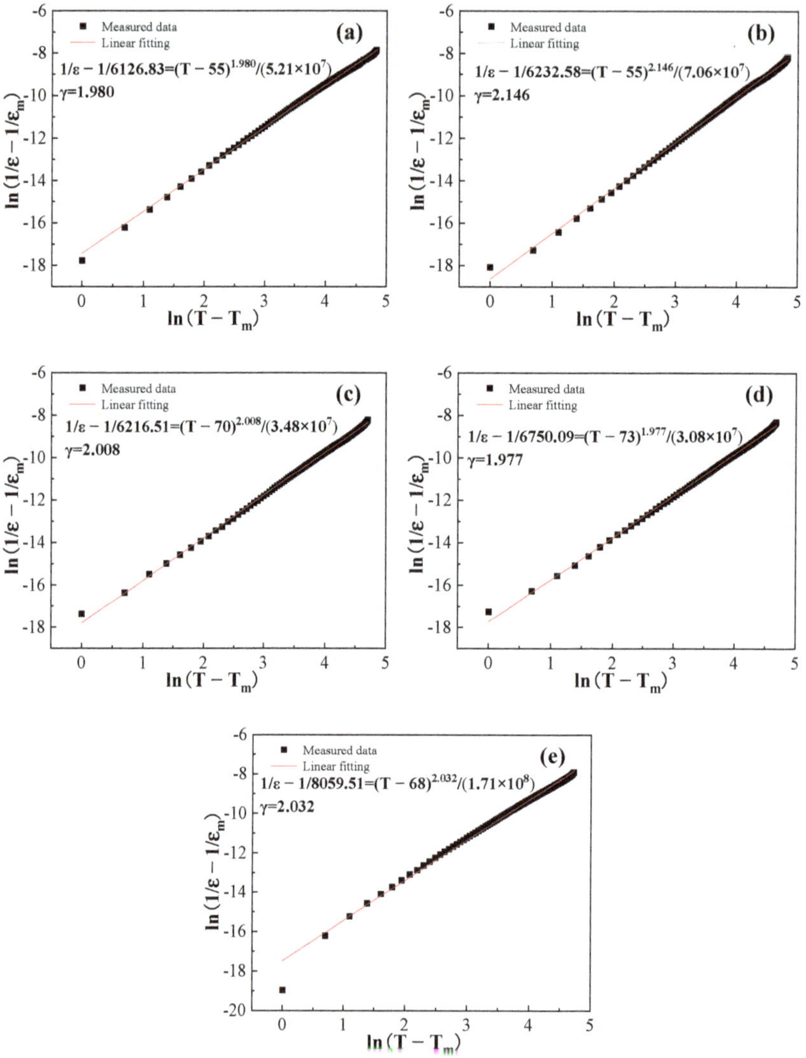

Figure 7. Dielectric exponential law fitting of the BCZTNb$_{0.001}$-0.025BiNZ ceramic thick films sintered at different sintering temperatures using 10 kHz data: (**a**) 1305 °C; (**b**) 1315 °C; (**c**) 1325 °C; (**d**) 1335 °C; (**e**) 1345 °C.

3.4. Conduction Mechanism

In order to investigate the electrical conductivity and resistivity properties of materials, the complex impedance spectra of the BCZTNb$_{0.001}$-0.025BiNZ ceramic thick films sintered at different sintering temperatures are measured and the Nyquist plots are shown in Figure 8, which is measured at elevated temperatures from 100 Hz to 1 MHz. It can be seen that all complex impedance spectra of all BCZTNb$_{0.001}$-0.025BiNZ films measured at different testing temperatures present a single semicircle arc. The semicircles at lower testing temperatures are incomplete since the external energy is insufficient to stimulate the conduction process of ions and point defects within the ceramic thick films [12,25]. As the testing temperature increases, there is a gradual decrease in the diameter of semicircles in all films, indicating that the resistivity of samples is decreasing and the conductiv-

ity is increasing [25,43]. The fitted equivalent circuits of complex impedance spectra at 430 °C of respective ceramic thick films are shown in the insets of Figure 8, where R is the circuit resistance and CPE is the constant phase element. The volume resistivity of ceramic materials is jointly determined by the grain boundary resistance and grain resistance [26,43], whereas the complex impedance spectra of all samples can be fitted well with a single R and a single CPE connected in parallel, and the fitted results are in good agreement with the experimental data. The above results indicate that the resistance of all ceramic thick films is dominated by grain resistance. The fitted R value decreases gradually as the sintering temperature increases since as the grain size increases, the grain resistance decreases [26], which is consistent with the conclusion in Figure 4.

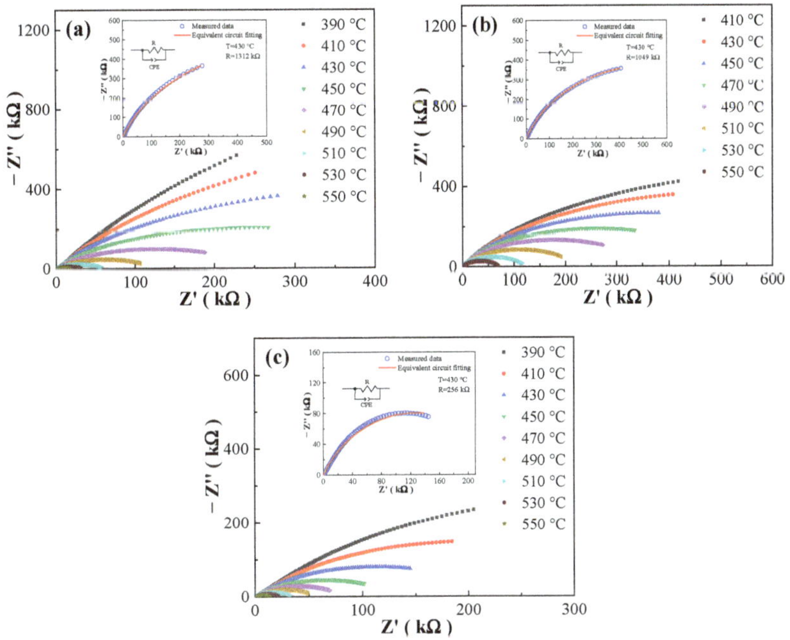

Figure 8. Z″ versus Z′ curves and the fitted equivalent circuits at 430 °C (insets) of the $BCZTNb_{0.001}$-0.025BiNZ ceramic thick films sintered at different sintering temperatures: (**a**) 1305 °C; (**b**) 1325 °C; (**c**) 1345 °C.

In accordance with the Arrhenius law (Equations (5) and (6)) [40,44], where σ_{dc} is the DC conductivity, τ is the relaxation time, k_b is the Boltzmann constant, T is the thermodynamic temperature, E_a is the activation energy, σ_0 and τ_0 are constant, the conductivity activation energy and the relaxation activation energy can be calculated separately by linearly fitting the relationship between DC conductivity (σ_{dc}) and relaxation time (τ) with temperature (T). Then, the conduction mechanism of the $BCZTNb_{0.001}$-0.025BiNZ ceramic thick films sintered at different sintering temperatures can be studied, as shown in Figure 9.

$$\sigma_{dc} = \sigma_0 \exp\left(-\frac{E_a}{Tk_b}\right) \quad (5)$$

$$\tau = \tau_0 \exp\left(-\frac{E_a}{Tk_b}\right) \quad (6)$$

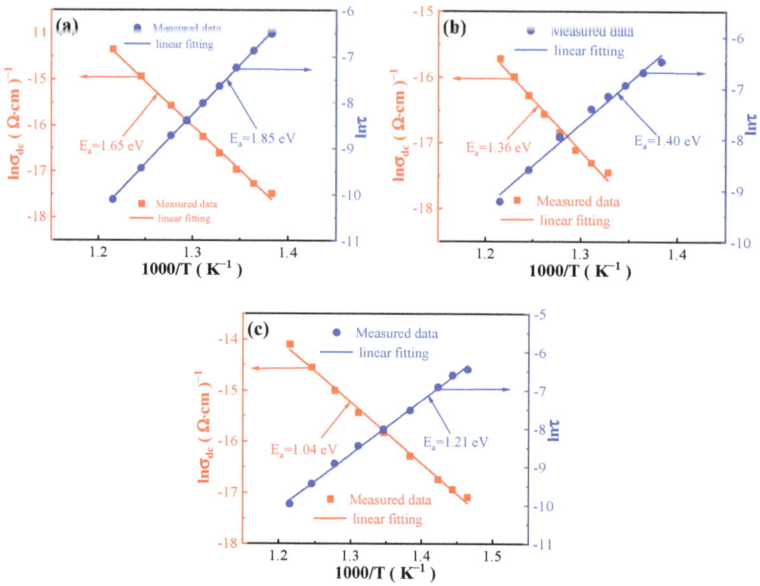

Figure 9. σ_{dc} and τ versus 1/T curves of the $BCZTNb_{0.001}$-0.025BiNZ ceramic thick films sintered at different sintering temperatures: (**a**) 1305 °C; (**b**) 1325 °C; (**c**) 1345 °C.

Based on the Arrhenius law linear fitting, the ceramic thick films sintered at different temperatures exhibit the conductivity activation energy of 1.65 eV (1305 °C), 1.36 eV (1325 °C) and 1.04 eV (1345 °C), while the relaxation activation energy is 1.85 eV (1305 °C), 1.40 eV (1325 °C) and 1.21 eV (1345 °C), respectively. Notably, the relaxation activation energy for each film exceeds the respective conductivity activation energy, which is caused by studying the thermal relaxation process from the perspectives of the dynamics of conductive rate and thermodynamics of conduction energy barrier [40]. The activation energy (E_a) for oxygen vacancy conduction is reported ranging from 0.5 eV to 2 eV [43], and the conductivity activation energy as well as the relaxation activation energy for all ceramic thick films fall within this range (Figure 9). Consequently, the conduction mechanism of all $BCZTNb_{0.001}$-0.025BiNZ ceramic thick films at high temperatures is dominated by oxygen vacancy conduction [43]. Such discussion confirms the speculated reason for the significant increase in dielectric loss, especially at low frequencies, under elevated test temperatures (Figure 6). The generation of oxygen vacancies is associated with the volatilization of certain metals (Na and Bi) and the heterovalent substitution of Bi^{3+} at the A-site and Nb^{5+} at the B-site of the perovskite structure [39] as illustrated in Equations (1)–(4) [40,41].

3.5. Ferroelectric and Field-Induced Strain Performance

The P-E hysteresis loops (60 kV/cm and 1 Hz) and bipolar S-E curves (60 kV/cm and 10 Hz) of the $BCZTNb_{0.001}$-0.025BiNZ ceramic thick films sintered at different temperatures are shown in Figure 10a,b. All ceramic thick films have relatively slender and symmetric P-E loops, and butterfly-like bipolar S-E curves with slight asymmetry, and the negative strain in the bipolar S-E curves can be ignored, which also indicates the relaxation characteristic of all $BCZTNb_{0.001}$-0.025BiNZ films [4,12,35]. There is a slight non-closure phenomenon in the P-E loops, which is related to the periodic polarization of space charge field generated during sintering due to sublimination of Na and Bi metals and heterovalent substitution of Bi^{3+} and Nb^{5+} [39,45]. The bipolar S-E curves also show slight asymmetry, which is related to the existence of the internal biased field established by space charge field [14]. With the increase in sintering temperature, the P-E loops and S-E curves tend to augment accompanied by the improvement of ferroelectricity and electrostrain.

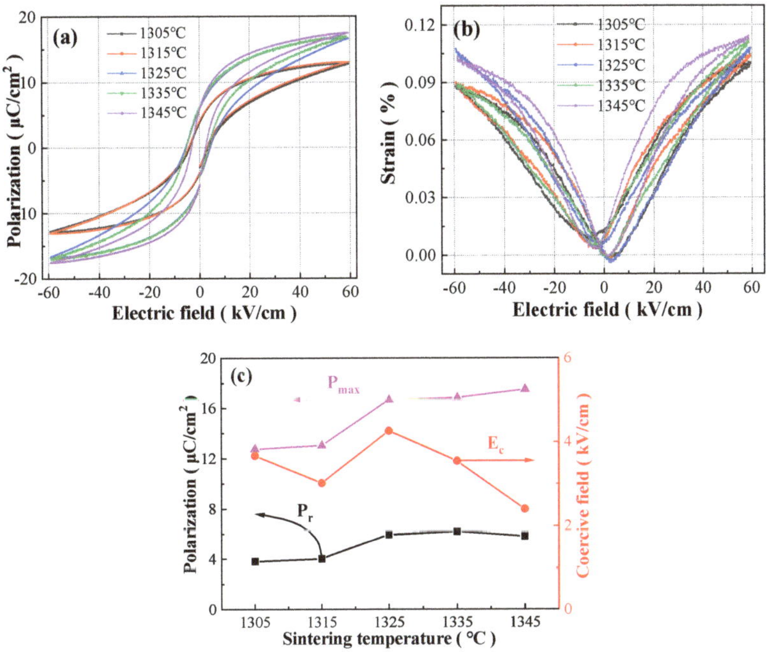

Figure 10. (**a**) P-E hysteresis loops at 60 kV/cm and 1 Hz and (**b**) bipolar S-E curves at 60 kV/cm and 10 Hz of the BCZTNb$_{0.001}$-0.025BiNZ ceramic thick films sintered at different sintering temperatures; (**c**) curves of polarization and coercive field with sintering temperature of the BCZTNb$_{0.001}$-0.025BiNZ ceramic thick films.

Figure 10c summarizes the variation in the maximum polarization (P_{max}), the remnant polarization (P_r) and the coercive field (E_c) of the BCZTNb$_{0.001}$-0.025BiNZ ceramic thick films with the sintering temperature. As the sintering temperature is raised from 1305 °C to 1345 °C, the P_{max} value of BCZTNb$_{0.001}$-0.025BiNZ films shows an upward trend, increasing from 12.76 μC/cm^2 to 17.51 μC/cm^2. On the other hand, P_r initially increases and then decreases, and the largest P_r value of 6.19 μC/cm^2 is obtained at a sintering temperature of 1335 °C. E_c fluctuates irregularly, presenting a decrease–increase–decrease trend, with the extremum minimum E_c value of 2.40 kV/cm at a sintering temperature of 1345 °C and the maximum E_c value of 4.26 kV/cm at a sintering temperature of 1325 °C. As a comparison, the P_r and E_c values of the 1450 °C sintered pure BCZT ceramics are ~15 μC/cm^2 and 1.68 kV/cm, respectively, reported by Ren et al. [7]. The E_c value of BCZTNb$_{0.001}$-0.025BiNZ ceramic films is rather small, which is easy for ferroelectric domain switching, and results in low hysteresis [19,20,33]. Comparatively speaking, the P_m, P_r and E_c values of 0.65BiFeO$_3$-0.35BaTiO$_3$ at 60 kV/cm were 38.2 μC/cm^2, 26 μC/cm^2 and 20.7 kV/cm, respectively, reported by Yuan et al. [14], and the P_m, P_r and E_c values of 0.67BiFe$_{99.75\%}$Sb$_{0.25\%}$O$_3$-0.33BaTiO$_3$ at 50 kV/cm were 33.23 μC/cm^2, 26.93 μC/cm^2 and ~28 kV/cm, respectively, reported by Kang et al. [19], whose E_c values are rather large. Additionally, the ferroelectric properties of ceramic thick films exhibit a certain improvement with the increase in sintering temperature, which is associated with the increase in grain size [29,31,33] and is consistent with the grain size variation trend observed in Figure 4.

Figure 11 shows the unipolar S-E curves (60 kV/cm and 10 Hz), and variations in the unipolar strain $S_{unipolar}$, the converse piezoelectric coefficient d_{33}^* ($d_{33}^* = S_{max}/E_{max}$ [46,47], S_{max} is the unipolar strain under the maximum electric field E_{max}) and the strain hysteresis Hys (Hys = $\Delta S/S_{max}$ [46,47], ΔS is the difference in strain at half maximum field) with sintering temperature of the BCZTNb$_{0.001}$-0.025BiNZ ceramic thick films sintered at different temperatures. The unipolar S-E curves of all ceramic thick films exhibit a similar slender shape, and the $S_{unipolar}$ value shows an upward trend, increasing from 0.099% (1305 °C) to 0.114% (1345 °C) as the sintering temperature increases. The corresponding d_{33}^* value also shows similar gradual increase with the elevation of sintering temperature, rising from 166.5 pm/V (1305 °C) to 191.3 pm/V (1345 °C). The increasing trend of $S_{unipolar}$ and d_{33}^* is also associated with the enlargement of grain size since the internal strain in the BCZT-based ceramics primarily originates from ferroelectric domain switching [33]. The larger grain size is more conducive to ferroelectric domain switching than the smaller grain size, so ceramic materials with larger grain size tend to have larger strain and d_{33}^* [19,31–33], which is consistent with the conclusion drawn from Figure 4.

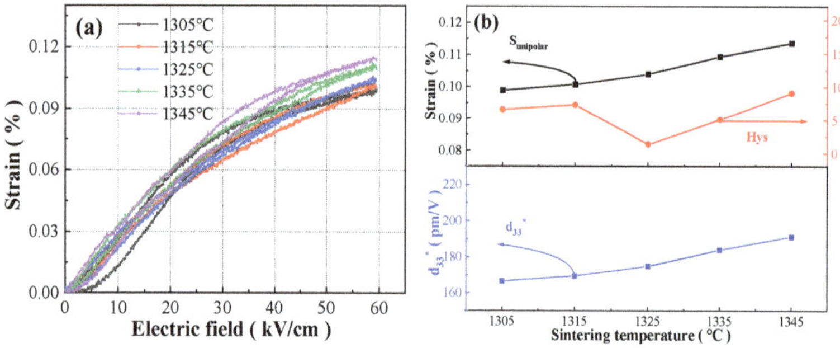

Figure 11. (**a**) Unipolar S-E curves at 60 kV/cm and 10 Hz of the $BCZTNb_{0.001}$-0.025BiNZ ceramic thick films sintered at different sintering temperatures; (**b**) curves of $S_{unipolar}$, d_{33}^* and Hys with sintering temperature.

Furthermore, all $BCZTNb_{0.001}$-0.025BiNZ ceramic thick films demonstrate very low Hys, each below 10%. The lowest Hys of 1.34% is achieved in the ceramic thick film sintered at 1325 °C accompanied by relatively high $S_{unipolar}$ of 0.104% at 60 kV/cm, which can effectively meet the requirement of piezoelectric actuators for high strain and low hysteresis. Such low Hys can be attributed to the disordered internal structure and enhanced random field in the $BCZTNb_{0.001}$-0.025BiNZ ceramic thick films resulting from the addition of a second component (BiNZ) and equivalent ion substitution, and A-site Bi^{3+} and B-site Nb^{5+} donor doping, which disrupts long-range ordered ferroelectric states and generates PNRs [12,13,19,21]. The intense response of PNRs to an external electric field can suppress strain hysteresis, thereby enabling $BCZTNb_{0.001}$-0.025BiNZ to maintain lower strain hysteresis [14,15,17].

Table 2 summarizes the $S_{unipolar}$ and Hys of the $BCZTNb_{0.001}$-0.025BiNZ ceramic thick film sintered at 1325 °C and several other lead-free perovskite ceramic materials reported recently [8,9,17,19,22,47]. It can be observed that the $BCZTNb_{0.001}$-0.025BiNZ ceramic thick film sintered at 1325 °C exhibits nearly zero Hys (1.34%) while it also maintains relatively large $S_{unipolar}$ (0.104%) compared with other lead-free ceramic materials, where ceramics that have larger strain also possess an unacceptably higher Hys, suggesting excellent development prospects in the field of actuators.

Table 2. Comparison of strain properties of the $BCZTNb_{0.001}$-0.025BiNZ ceramic thick film sintered at 1325 °C with some lead-free ceramics reported in the literature.

Compounds	$S_{unipolar}$ (%)	Hys (%)	Electric Field (kV/cm)	Ref.
$Ba_{0.6}Sr_{0.4}TiO_3$	0.07	4	60	[8]
$0.91(K_{0.48}Na_{0.52})Nb_{0.965}Sb_{0.035}$-$0.03Bi_{0.5}(K_{0.18}Na_{0.82})_{0.5}ZrO_3$-$0.06BaZrO_3$	0.05	8	40	[9]

Table 2. *Cont.*

Compounds	$S_{unipolar}$ (%)	Hys (%)	Electric Field (kV/cm)	Ref.
$0.94BaTiO_3$-$0.06KNbO_3$	0.10	3	80	[17]
$0.67BiFe_{99.75\%}Sb_{0.25\%}O_3$-$0.33BaTiO_3$	0.168	49	50	[19]
$0.70Bi_{1.03}FeO_3$-$0.30Ba_{0.985}La_{0.015}TiO_3$	0.188	20	55	[22]
$0.91Bi_{0.5}Na_{0.5}TiO_3$-$0.07Ba(Zr_{0.055}Ti_{0.945})O_3$-$0.02(K_{0.5}Na_{0.5})NbO_3$	0.31	33.5	65	[47]
$BCZTNb_{0.001}$-0.025BiNZ	0.104	1.34	60	This work

4. Outlook

Presently, among the relatively effective approaches to improve electrostrain performance is the formation of PNRs within materials through doping cations or adding a second component [13,15,17]. In this article, the 1325 °C sintered $BCZTNb_{0.001}$-0.025BiNZ ceramic thick film has attained an extremely low strain hysteresis (1.34%) at 60 kV/cm by generating PNRs; however, the unipolar strain (0.104%) still requires further enhancement. It is reported that the size and activity of PNRs exert an influence on the strain performance of ceramics. Small-sized and highly active PNRs will enable the material to achieve ultra-low strain hysteresis, but unfortunately, the corresponding strain will also decline [15]. Controlling the ferroelectric relaxor transition temperature is one of the known methodologies [38]. In the future, how to effectively regulate the size and activity of PNRs will constitute the key to attaining materials with extremely low hysteresis and high strain.

5. Conclusions

In this work, the $BCZTNb_{0.001}$-0.025BiNZ ceramic thick films prepared by scraping process exhibit a pure perovskite structure, typical relaxation characteristic and high-temperature conduction mechanism of oxygen vacancies at different sintering temperatures. The scraping process simplifies the process of preparing thinner ceramic materials while significantly improving the external electric field resistance of ceramics. Due to the addition of a second component BiNZ with a low melting point, the ceramic thick film can be densely sintered at a low temperature of approximately 1325 °C. The grain size, dielectric, ferroelectric and field-induced strain properties of the $BCZTNb_{0.001}$-0.025BiNZ ceramic thick films increase with elevating the sintering temperature. The 1325 °C sintered ceramic thick film has the lowest hysteresis (1.34%) and relatively high unipolar strain (0.104%) at 60 kV/cm, solving the problem that low hysteresis and large strain cannot coexist in the material, and showing good application prospects in the actuator field.

Author Contributions: Conceptualization, B.F.; supervision, B.F. and J.D.; formal analysis, Y.Z. and B.F.; investigation, Y.Z. and S.Z.; data curation, Y.Z. and X.L.; writing—original draft, Y.Z.; writing—review and editing, B.F.; funding acquisition, B.F., X.L. and S.Z.; project administration, B.F. and J.D. All authors have read and agreed to the published version of the manuscript.

Funding: This work was supported by the Top-notch Academic Programs Project of Jiangsu Higher Education Institutions and the Priority Academic Program Development of Jiangsu Higher Education Institutions.

Institutional Review Board Statement: Not applicable.

Informed Consent Statement: Not applicable.

Data Availability Statement: All data that support the findings of this study are included within the article, or available from the corresponding author upon request.

Conflicts of Interest: The authors declare that they have no known competing financial interests or personal relationships that could have appeared to influence the work reported in this paper.

Nomenclature

Amm2	an orthorhombic phase space group
P4mm	a tetragonal phase space group
R3m	a rhombohedral phase space group
BCZT	$(Ba_{0.85}Ca_{0.15})(Zr_{0.1}Ti_{0.9})O_3$
BCZT-xBiNZ	$(1 - x)(Ba_{0.85}Ca_{0.15})(Zr_{0.1}Ti_{0.9})O_3$-$x(Bi_{0.5}Na_{0.5})ZrO_3$
$BCZTNb_x$-0.025BiNZ	$0.975(Ba_{0.85}Ca_{0.15})[(Zr_{0.1}Ti_{0.9})_{1-x}Nb_x]O_3$-$0.025(Bi_{0.5}Na_{0.5})ZrO_3$
$BCZTNb_{0.001}$-0.025BiNZ	$0.975(Ba_{0.85}Ca_{0.15})[(Zr_{0.1}Ti_{0.9})_{0.999}Nb_{0.001}]O_3$-$0.025(Bi_{0.5}Na_{0.5})ZrO_3$
BiNZ	$(Bi_{0.5}Na_{0.5})ZrO_3$
CPE	constant phase element
C'	constant
d_{33}	piezoelectric constant
d_{33}^*	converse piezoelectric coefficient
E_a	activation energy
E_c	coercive field
E_{max}	maximum electric field
Hys	strain hysteresis
k_b	Boltzmann constant
P_m	maximum polarization
P-E	polarization-electric field hysteresis loop
PNRs	polar nanoregions
P_r	remnant polarization
R	circuit resistance
R_p	Rietveld refinement fitting reliability parameter
R_{wp}	Rietveld refinement fitting reliability parameter
χ^2	Rietveld refinement fitting reliability parameter
SEM	scanning electron microscope
S-E	strain-electric field loop
S_{max}	unipolar strain under the maximum electric field
$S_{unipolar}$	unipolar strain
T	temperature
T_m	the temperature of maximum dielectric constant
XRD	X-ray diffraction
Z'	impedance real part
Z''	impedance imaginary part
ε_m	maximum dielectric constant
γ	dispersion factor
σ_{dc}	DC conductivity
σ_0	constant
τ	relaxation time
τ_0	constant
ΔS	difference in strain at half maximum field

References

1. Limpichaipanit, A.; Ngamjarurojana, A. Strain characteristics of PLZT-based ceramics for actuator applications. *Actuators* **2023**, *12*, 74. [CrossRef]
2. Wang, J.; Wang, S.; Li, X.; Li, L.; Liu, Z.; Zhang, J.; Wang, Y. High piezoelectricity and low strain hysteresis in PMN-PT-based piezoelectric ceramics. *J. Adv. Ceram.* **2023**, *12*, 792–802. [CrossRef]
3. Liu, D.; Zeng, J.; Zheng, L.; Man, Z.; Ruan, X.; Shi, X.; Li, G. High piezoelectric property and low electric field-strain hysteresis of $BiAlO_3$-doped PZT ceramics. *J. Inorg. Mater.* **2022**, *37*, 1365–1370. [CrossRef]
4. Yu, D.; Zhou, C.; Chen, J.; Li, Q.; Li, L.; Yuan, C.; Xu, J.; Rao, G. Large electrostriction-like strain by tailoring relaxor degree in BNT-based ceramics. *J. Mater. Sci. Mater. Electron.* **2023**, *34*, 1153. [CrossRef]
5. Yuan, H.; Li, L.; Hong, H.; Ying, Z.; Zheng, X.; Zhang, L.; Wen, F.; Xu, Z.; Wu, W.; Wang, G. Low sintering temperature, large strain and reduced strain hysteresis of $BiFeO_3$-$BaTiO_3$ ceramics for piezoelectric multilayer actuator applications. *Ceram. Int.* **2021**, *47*, 31349–31356. [CrossRef]
6. Yao, K.; Zhou, C.; Li, Q.; Xiao, Z.; Yuan, C.; Xu, J.; Chen, G.; Rao, G. Large electrostrictive coefficient with optimized electro-strain in BNT-based ceramics with ergodic state. *Mater. Sci. Eng. B* **2022**, *283*, 115828. [CrossRef]

7. Liu, W.; Ren, X. Large piezoelectric effect in Pb-free ceramics. *Phys. Rev. Lett.* **2009**, *103*, 257602. [CrossRef]
8. Song, M.; Sun, X.; Li, Q.; Qian, H.; Liu, Y.; Lyu, Y. Enhanced electrostrictive coefficient and suppressive hysteresis in lead-free $Ba_{(1-x)}Sr_xTiO_3$ piezoelectric ceramics with high strain. *Crystals* **2021**, *11*, 555. [CrossRef]
9. Zhang, J.; Liu, Z.; Zhang, T.; Liu, Y.; Lyu, Y. High strain response and low hysteresis in $BaZrO_3$-modified KNN-based lead-free relaxor ceramics. *J. Mater. Sci. Mater. Electron.* **2021**, *32*, 16715–16725. [CrossRef]
10. Zhao, N.; Fan, H.; Li, C.; Huang, F.; Cao, J.; Li, Z. Large strain and low hysteresis in $(0.64\text{-}x)Bi_{0.5}Na_{0.5}TiO_3\text{-}0.36Sr_{0.7}Bi_{0.2}TiO_3\text{-}x(K_{0.5}La_{0.5})(Ti_{0.9}Zr_{0.1})O_3$ piezoceramics. *Ceram. Int.* **2021**, *47*, 30399–30405. [CrossRef]
11. Liu, H.; Hao, Y.; Yang, Z.; Feng, T.; Su, B.; Zhang, X.; Xue, M.; Zhang, B.; Li, J. Exceptional electrostrain with minimal hysteresis and superior temperature stability under low electric field in KNN-based lead-free piezoceramics. *J. Adv. Ceram.* **2024**, *13*, 364–372. [CrossRef]
12. Du, Y.; Qi, Y.; Hu, L.; Wan, H.; Wang, H.; Zhang, Q.; Zhang, G.; Jiang, S.; Shen, M.; Chen, Y. $Na_{0.5}Bi_{0.5}TiO_3$-based ferroelectric relaxor with high thermal stability and anti-fatigue for charge-discharge properties. *Ceram. Int.* **2021**, *47*, 15834–15842. [CrossRef]
13. Zeng, F.; Zhang, J.; Zhou, C.; Jiang, L.; Guo, H.; Chen, Y.; Lu, W.; Zhang, H.; Cai, W.; Hu, Y.; et al. Enhanced electric field-induced strain properties in lead-free BF-BT-based piezoceramics by local structure inhomogeneity. *ACS Sustain. Chem. Eng.* **2022**, *10*, 1277–1286.
14. Yuan, J.; Ruan, T.; Li, Q.; Liu, Y.; Lyu, Y. Large strain and reduced hysteresis of $BiFeO_3\text{-}xBaTiO_3$ lead-free ceramic via multiphase coexistence and relaxor behaviors. *Ceram. Int.* **2022**, *48*, 26335–26341. [CrossRef]
15. Hou, L.; Zhou, C.; Li, Q.; Li, R.; Yuan, C.; Xu, J.; Rao, G. Giant strain with ultra-low hysteresis by tailoring relaxor temperature and PNRs dynamic in BNT-based lead-free piezoelectric ceramics. *Ceram. Int.* **2022**, *48*, 13125–13133. [CrossRef]
16. Tian, A.; Yu, Z.; Xie, A.; Fu, J.; Zuo, R. Multiple local polarization configurations in NKN-based relaxor ferroelectric ceramic with high electrostrictive-like strain. *J. Eur. Ceram. Soc.* **2024**, *44*, 7614–7622. [CrossRef]
17. Huang, Y.; Zhao, C.; Wu, B.; Wu, J. Multifunctional $BaTiO_3$-based relaxor ferroelectrics toward excellent energy storage performance and electrostrictive strain benefiting from crossover region. *ACS Appl. Mater. Interfaces* **2020**, *12*, 23885–23895. [CrossRef]
18. Slouka, C.; Kainz, T.; Navickas, E.; Walch, G.; Hutter, H.; Reichmann, K.; Fleig, J. The effect of acceptor and donor doping on oxygen vacancy concentrations in lead zirconate titanate (PZT). *Materials* **2016**, *9*, 945. [CrossRef] [PubMed]
19. Kang, F.; Zhang, L.; He, L.; Sun, Q.; Wang, Z.; Kang, R.; Mao, P.; Zhu, C.; Wang, J. Enhanced electric field induced strain in B-site Sb doped $BiFeO_3\text{-}BaTiO_3$ lead free ceramics. *J. Alloys Compd.* **2021**, *864*, 158917. [CrossRef]
20. Wang, D.; Cao, M.; Yuan, J.; Zhao, Q.; Li, H.; Zhang, D.; Agathopoulos, S. Enhanced piezoelectric and ferroelectric properties of Nb_2O_5 modified lead zirconate titanate-based composites. *J. Am. Ceram. Soc.* **2011**, *94*, 647–650. [CrossRef]
21. Habib, M.; Akram, F.; Ahmad, P.; Al-Harbi, F.F.; Din, I.U.; Iqbal, Q.; Ahmed, T.; Khan, S.A.; Hussain, A.; Song, T.K.; et al. Donor multiple effects on the ferroelectric and piezoelectric performance of lead-free $BiFeO_3\text{-}BaTiO_3$ ceramics. *Mater. Lett.* **2022**, *315*, 131950. [CrossRef]
22. Habib, M.; Lee, M.H.; Kim, D.J.; Choi, H.I.; Kim, M.H.; Kim, W.J.; Song, T.K.; Choi, K.S. Enhanced piezoelectric performance of donor La^{3+}-doped $BiFeO_3\text{-}BaTiO_3$ lead-free piezoceramics. *Ceram. Int.* **2020**, *46*, 7074–7080. [CrossRef]
23. Li, H.; Fang, B.; Zhang, S.; Lu, X.; Ding, J. Low-temperature sintering PZT-based ceramics for extreme condition application. *J. Mater. Sci. Mater. Electron.* **2023**, *34*, 1970. [CrossRef]
24. Wang, H.; Hu, Q.; Liu, X.; Zheng, Q.; Jiang, N.; Yang, Y.; Kwok, K.W.; Xu, C.; Lin, D. A high-tolerance BNT-based ceramic with excellent energy storage properties and fatigue/frequency/thermal stability. *Ceram. Int.* **2019**, *45*, 23233–23240. [CrossRef]
25. Sun, J.; Yan, G.; Fang, B.; Zhang, S.; Lu, X.; Ding, J. Improving energy storage performance of BLLMT ceramic by doping BZT combining with defect engineering and film scraping process. *J. Alloys Compd.* **2024**, *971*, 172708. [CrossRef]
26. Yan, G.; Xu, L.; Fang, B.; Zhang, S.; Lu, X.; Zhao, X.; Ding, J. Achieving high pulse charge-discharge energy storage properties and temperature stability of $(Ba_{0.98\text{-}x}Li_{0.02}La_x)(Mg_{0.04}Ti_{0.96})O_3$ lead-free ceramics via bandgap and defect engineering. *Chem. Eng. J.* **2022**, *450*, 137814. [CrossRef]
27. Yan, F.; Huang, K.; Jiang, T.; Zhou, X.; Shi, Y.; Ge, G.; Shen, B.; Zhai, J. Significantly enhanced energy storage density and efficiency of BNT-based perovskite ceramics via A-site defect engineering. *Energy Storage Mater.* **2020**, *30*, 392–400. [CrossRef]
28. Beuerlein, M.A.; Kumar, N.; Usher, T.M.; Brown-Shaklee, H.J.; Raengthon, N.; Reaney, I.M.; Cann, D.P.; Jones, J.L., Brennecka, G.L. Current understanding of structure-processing-property relationships in $BaTiO_3\text{-}Bi(M)O_3$ dielectrics. *J. Am. Ceram. Soc.* **2016**, *99*, 2849–2870. [CrossRef]
29. He, X.; Fang, B.; Zhang, S.; Lu, X.; Ding, J. Preparation and properties of Nd-doped BCTH lead-free ceramics by solid-phase twin crystal method. *Curr. Appl. Phys.* **2022**, *38*, 30–39. [CrossRef]
30. Tsuji, K.; Ndayishimiye, A.; Lowum, S.; Floyd, R.; Wang, K.; Wetherington, M.; Maria, J.P.; Randall, C.A. Single step densification of high permittivity $BaTiO_3$ ceramics at 300 °C. *J. Eur. Ceram. Soc.* **2020**, *40*, 1280–1284. [CrossRef]
31. Bijalwan, V.; Hughes, H.; Pooladvand, H.; Tofel, P.; Nan, B.; Holcman, V.; Bai, Y.; Button, T.W. The effect of sintering temperature on the microstructure and functional properties of BCZT-$xCeO_2$ lead free ceramics. *Mater. Res. Bull.* **2019**, *114*, 121–129. [CrossRef]
32. Kuhfuss, M.; Maier, J.G.; Hall, D.A.; Xie, B.; Kleppe, A.K.; Martin, A.; Kakimoto, K.I.; Khansur, N.H.; Webber, K.G. In situ electric field-dependent structural changes in $(Ba,Ca)(Zr,Ti)O_3$ with varying grain size. *J. Appl. Phys.* **2024**, *135*, 174101. [CrossRef]

33. Cai, W.; Zhang, Q.; Zhou, C.; Gao, R.; Zhang, S.; Li, Z.; Xu, R.; Chen, G.; Deng, X.; Wang, Z.; et al. Synergistic effect of grain size and phase boundary on energy storage performance and electric properties of BCZT ceramics. *J. Mater. Sci. Mater. Electron.* **2020**, *31*, 9167–9175. [CrossRef]
34. Dai, Z.; Xie, J.; Chen, Z.; Zhou, S.; Liu, J.; Liu, W.; Xi, Z.; Ren, X. Improved energy storage density and efficiency of (1-x)$Ba_{0.85}Ca_{0.15}Zr_{0.1}Ti_{0.9}O_3$-x$BiMg_{2/3}Nb_{1/3}O_3$ lead-free ceramics. *Chem. Eng. J.* **2021**, *410*, 128341. [CrossRef]
35. Deng, A.; Wu, J. Optimized strain properties with small hysteresis in BNT-based ceramics with ergodic relaxor state. *J. Eur. Ceram. Soc.* **2021**, *41*, 5147–5154. [CrossRef]
36. German, R.M.; Suri, P.; Park, S.J. Liquid phase sintering. *J. Mater. Sci.* **2009**, *44*, 1–39. [CrossRef]
37. Jaiban, P.; Theethuan, T.; Khumtrong, S.; Lokakaew, S.; Kongchana, P.; Watcharapasorn, A. The effects of A-site doping on crystal structure, microstructure and electrical properties of $(Ba_{0.85}Ca_{0.15})$ $Zr_{0.1}Ti_{0.9}O_3$ ceramics. *Chiang Mai J. Sci.* **2022**, *49*, 1150–1163. [CrossRef]
38. Wang, X.; Liu, X.; Xue, H.; Yin, J.; Wu, J. Temperature-independent large strain with small hysteresis in Sb-modified BNT-based lead-free ceramics. *J. Am. Ceram. Soc.* **2022**, *105*, 2116–2127. [CrossRef]
39. Wang, Y.; Jie, W.; Yang, C.; Wei, X.; Hao, J. Colossal permittivity materials as superior dielectrics for diverse applications. *Adv. Funct. Mater.* **2019**, *29*, 1808118. [CrossRef]
40. Li, W.; Fang, B.; Lu, X.; Zhang, S.; Ding, J. $(Bi_{0.5}Na_{0.5})(Zr_{0.97}W_{0.01}Fe_{0.02})O_3$ doping effects on KNN-based lead-free ceramics prepared via two-step sintering technique. *Int. J. Appl. Ceram. Technol.* **2024**, *21*, 515–524. [CrossRef]
41. Jaiban, P.; Theethuan, T.; Khumtrong, S.; Lokakaew, S.; Watcharapasorn, A. The effects of donor (Nb^{5+}) and acceptor (Cu^{2+}, Zn^{2+}, Mn^{2+}, Mg^{2+}) doping at B-site on crystal structure, microstructure, and electrical properties of $(Ba_{0.85}Ca_{0.15})Zr_{0.1}Ti_{0.9}O_3$ ceramics. *J. Alloys Compd.* **2022**, *899*, 162909. [CrossRef]
42. Li, L.; Roncal-Herrero, T.; Harrington, J.; Milne, S.J.; Brown, A.P.; Dean, J.S.; Sinclair, D.C. Anomalous grain boundary conduction in $BiScO_3$-$BaTiO_3$ high temperature dielectrics. *Acta Mater.* **2021**, *216*, 117136. [CrossRef]
43. Wang, L.; Bai, W.; Zhao, X.; Ding, Y.; Wu, S.; Zheng, P.; Li, P.; Zhai, J. Influences of rare earth site engineering on piezoelectric and electromechanical response of $(Ba_{0.85}Ca_{0.15})(Zr_{0.1}Ti_{0.9})O_3$ lead-free ceramics. *J. Mater. Sci. Mater. Electron.* **2020**, *31*, 6560–6573. [CrossRef]
44. Sareecha, N. Marked influence of low Bi doping levels on the structural and thermal transport properties of nonstoichiometric 0.98 $PbTiO_3$ ceramics. *Mater. Chem. Phys.* **2019**, *229*, 124–129. [CrossRef]
45. Tang, X.G.; Chew, K.H.; Chan, H.L. Diffuse phase transition and dielectric tunability of $Ba(Zr_yTi_{1-y})O_3$ relaxor ferroelectric ceramics. *Acta Mater.* **2004**, *52*, 5177–5183. [CrossRef]
46. Zhang, R.; Wang, P.; Guo, Q.; Long, J.; Huang, T.; Yao, Z.; Cao, M.; Liu, H.; Hao, H. High electric field-induced strain properties in La-doped $Pb(In_{1/2}Nb_{1/2})$ O_3-$PbZrO_3$-$PbTiO_3$ relaxor ferroelectric ceramics. *Ceram. Int.* **2023**, *49*, 31582–31590. [CrossRef]
47. Dong, S.; Guo, F.; Long, W.; Li, X.; Fang, P.; Dai, Z.; Xi, Z. Enhanced piezoelectric and electric field induced stain response in $Bi_{0.5}Na_{0.5}TiO_3$-$Ba(Zr_{0.055}Ti_{0.945})O_3$ ceramics modified by $(K_{0.5}Na_{0.5})NbO_3$. *J. Mater. Sci. Mater. Electron.* **2020**, *31*, 13979–13986. [CrossRef]

MDPI AG
Grosspeteranlage 5
4052 Basel
Switzerland
Tel.: +41 61 683 77 34

Materials Editorial Office
E-mail: materials@mdpi.com
www.mdpi.com/journal/materials

www.ingramcontent.com/pod-product-compliance
Lightning Source LLC
LaVergne TN
LVHW070447061126
851858LV00011B/183